Issued under the authority of the Home Office
(Fire and Emergency Plannin[

Fire Servic[

Volume 2
Fire Service Operations

Incident Command

HM Fire Service Inspectorate Publications Section
London: The Stationery Office

ISBN 0 11 341191 X

Cover photograph and half-title page photograph:
West Yorkshire Fire Service

Printed in the United Kingdom for The Stationery Office
J72789 C50 2/99 5673

Preface

This manual brings together the lessons learned from decades of incident command in the fire service and describes the principles upon which systems of incident command should be based. It does not seek to impose prescriptive methods of applying such systems, but concentrates instead on establishing the key components of an effective system to support competent incident command performance.

The adoption of a standard system for the command of fires and emergency incidents for the fire service reflects the need to ensure a safe system of work at all operational emergency incidents. Such a system also provides a consistent and transparent framework for training in command against which performance can be assessed.

In recognition of the fact that incident commanders operate in a dynamic and uncertain environment, the system also includes an explanation of the process of dynamic risk assessment.

The Incident Command System is built upon, and develops, existing good practice. The system is provided for the use and guidance of fire brigades in assisting them to deliver the core operational service – managing incidents – and managing them safely.

It features the following elements:

(i) a standard structure for organising resources on the fireground which takes account of the dynamic nature of an emergency incident, assists Incident Commanders' decision making and encourages effective communication and delegation;

(ii) a process of dynamic risk assessment which ensures that Incident Commanders retain safety considerations at the forefront of their command decision-making process and the method of demonstrating this;

(iii) a system of structured support for the Incident Commander which ensures that communications are exercised within a manageable span of control;

(iv) a national framework which describes the tasks and the performance standards expected of commanders at emergency incidents;

(v) a consistent methodology for the procedures and practices of operational command which will contribute to safer systems of work both in training and at operational incidents, especially for fire-fighters attending 'cross-border' incidents and external training centres;

(vi) a recognition of the role of performance management and review which, in the operational context, centres particularly on post-incident debrief procedures.

The safety of fire-fighters at operational incidents and during realistic training is a critical area of responsibility for fire service commanders and managers.

Although it is not possible to define precise procedures for the very wide range of emegency incidents attended by brigades, there are essential features of organisation and command which can be applied for their better management and conclusion.

The frequent need for fire-fighters from one brigade to work under the command of officers from another, the need for fire-fighters to work with other emergency services and the need for fire-fighters and commanders at all levels to be trained at national centres, all highlight the need for systems of work which operate commonly to national standards. The contents of this manual were developed to meet these requirements.

The systems, procedures and practices described in the following pages were developed under the 'Safe Person Concept' with the intention of setting out good practice and offering supporting advice and guidance in those areas of operational activity considered to be critical to the health and safety of fire service personnel.

Under the 'Safe Person Concept' umbrella, project groups, comprising representatives of all key fire service interests in the operational field, have been working under the leadership of HM Fire Service Inspectorate to deal with specific issues which directly impinge on the safety of fire-fighters.

The safe and competent command of operational incidents is one of these and the Home Office is indebted to members of the 'Incident Command' Project Group whose efforts have ensured the production of this manual.

Incident Command

Contents

Incident Command

Chapter 1 – Introduction to the Incident Command System

THE
INCIDENT COMMAND
SYSTEM

ORGANISATION
ON THE
INCIDENT
GROUND

COMMAND
COMPETENCE

DYNAMIC
RISK
ASSESSMENT

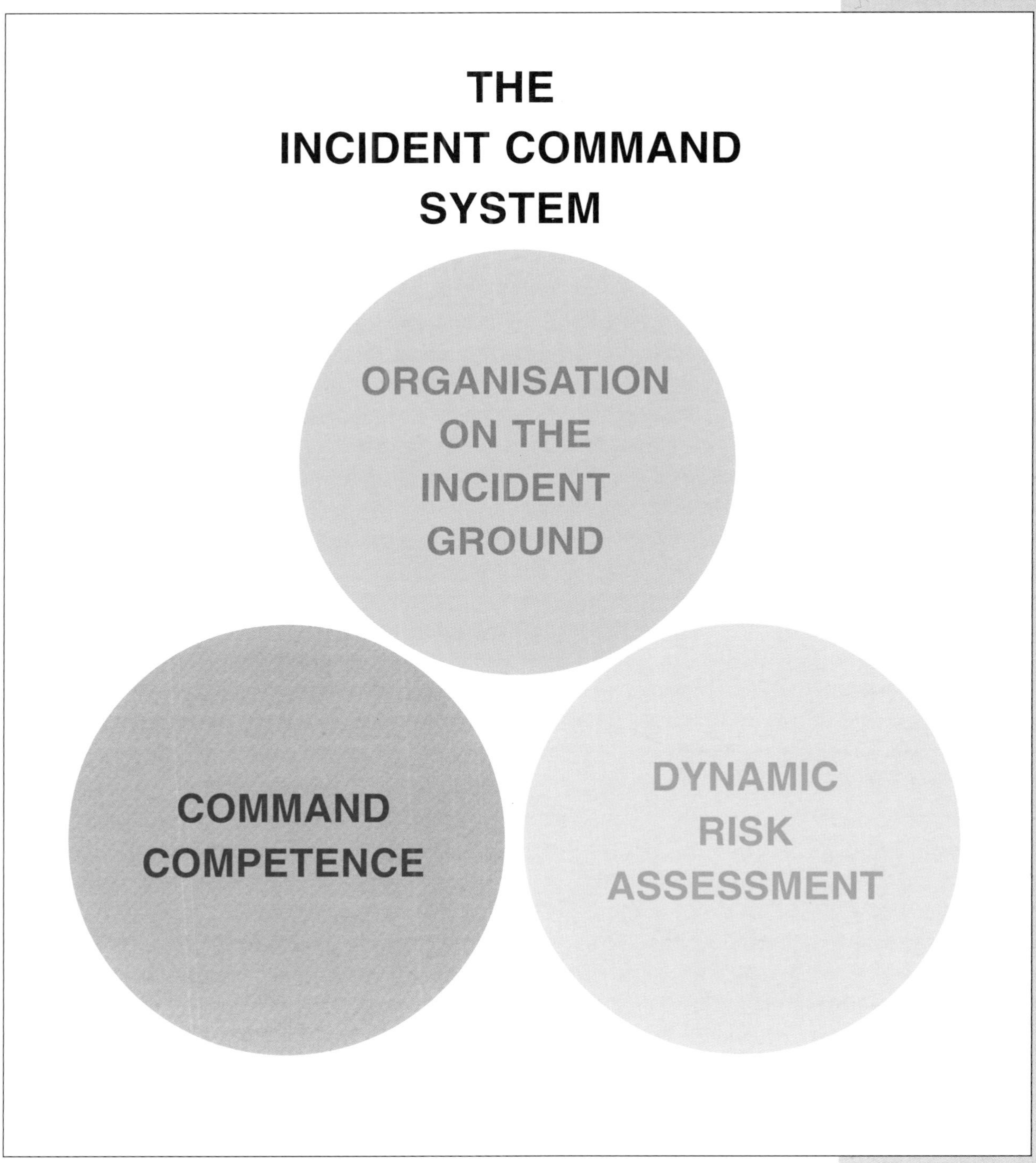

1.1 Incident Command System in Context

An Incident Command System cannot exist in isolation. The subsequent sections of this Manual address, principally, the design of systems for incident command and the training and assessment of individuals and teams to operate those systems safely and effectively. There are, however, a number of other critical factors which support the incident command function and which a brigade will need to consider in integrating the function within its management system. The development of an Incident Command System should be seen as part of a brigade's overall organisational system for managing risk. The approach advocated by the Health & Safety Executive in HSG65 to the design of organisational structures and processes for managing safely and successfully, provide a useful framework for this.

1.1.1 Policy

A brigade should have a clear and coherent policy that sets out the approach to delivering effective incident command.

1.1.2 Organising

The arrangements by which the Incident Command System is delivered and supported should be defined clearly. This will allow all involved to understand the brigade's approach and objectives in relation to the command function. Examples of issues to be addressed would include the provision of vehicle/equipment availability, mobilising arrangements and training facilities, resources and programmes.

1.1.3 Planning

There should be a planned approach to the development and implementation of the incident command function, the aim of which should be to minimise and mitigate risks. Risk assessment should be used to identify priorities for the development of the Incident Command System and to set objectives to eliminate hazards and reduce or control risk. Issues relating to operational preplanning would feature here.

1.1.4 Measuring

There should be agreed and documented standards of performance in incident command and a system in place to measure performance and identify areas for improvement.

1.1.5 Auditing & Reviewing Performance

A system should be adopted which enables the brigade to undertake reviews of incident command performance to ensure that all relevant experience can be captured and lessons learned.

1.2 The Key Elements of Incident Command

Successful incident command requires certain key features.

1.2.1 The Incident Commander

The *fire service incident commander* at an operational incident has the right to exercise authority over fire service resources on the incident ground.

The Incident Commander has much to consider when dealing with an emergency and this will intensify with its scale and duration. Clearly, no officer can be expected to remember everything, so the system of incident command described in this manual will provide operational and managerial prompts to reinforce those given by the incident itself and the personnel in support roles.

The Incident Commander must ensure that adequate resources are available and that arrangements have been made to control them. At larger incidents these will normally be made the responsibility of supporting officers in the command structure.

Good communication between personnel is essential throughout the incident but especially at the time of the handing over of command which can result in confusion if it is not done properly. The accumulated knowledge of the site, the incident, the risks and the actions taken so far need to be communicated, in an easily assimilated form, to the officer taking over.

An Incident Commander should be prepared to brief a more senior officer at any time so that he/she can make a decision whether or not to assume command. If the senior officer decides to take command he/she is to inform the current Incident Commander by stating "I am taking over". Having assumed command the senior officer may elect to retain the previous commander in the command structure to give assistance.

It is the duty of officers being relieved to give the senior officer all the relevant information they possess concerning the incident. Handover of command to more junior officers as the incident is being reduced in size must be equally thorough.

1.2.2 Strategy, Tactics & Operations

Strategy, tactics and operations are the descriptions given to the different managerial levels of fire service activity on the incident ground. These terms need to be understood in the context of the incident command structure.

***Strategy* is the planning and directing of the organisation in order to meet its overall objectives.**

For fire brigade operations such objectives would be likely to include:

- Saving and protecting those in danger.
- Ensuring the safety of operational personnel.
- Protecting property.
- Protecting the environment.

The *strategic* level of responsibility includes the formulation of systems to manage the risks of certain incident types in seeking to achieve these objectives. Such systems should be considered best practice for safe and effective work on the incident ground.

As each operational incident may differ, Incident Commanders may need to adapt the strategic plans and systems in the light of the risks presented by the incident and the resources available to deal with those risks.

***Tactics* can be summarised as the deployment of personnel and equipment on the incident ground to achieve the strategic aims of the Incident Commander.**

These will almost invariably be based on approved operational procedures.

***Operations* can best be described as tasks that are carried out on the incident ground, using prescribed techniques and procedures in accordance with the tactical plan.**

At the smallest incidents all three levels of command decision making will be the responsibility of one individual, likely to be the first arriving Crew Commander, who will be very much concerned with the tactics and operational tasks in the initial stages, delegating responsibility for the operational level if sufficient resources are available. At larger incidents the team of officers responsible for the various command functions will be organised by the Incident Commander to discharge the tactical and operational tasks, while the Incident Commander retains overall command.

1.2.3 Resources & Control

The Incident Commander is responsible for securing and controlling resources on the incident ground. The assessment of resources will include:

- Appliances
- Personnel
- Equipment
- Firefighting media
- Consumables (e.g. fuel, BA cylinders)

The degree of control an Incident Commander will need to maintain will depend, in part, on the size and demands of the incident. At larger incidents specific areas of resource control may be delegated to appointed officers. Such areas may include:

- Firefighting
- Command support
- Marshalling
- Logistics

- Decontamination
- Water
- Foam
- BA

and may be designated as **'sectors'** for the purposes of control and identification.

When considering control of the incident ground, the incident commander will, in particular, consider the need to maintain the safety of fire-fighters, the public and members of other emergency services attending.

Cordons may be employed as an effective method of controlling resources and maintaining safety. There are two types, inner and outer cordons.

Inner cordons to control the immediate scene of operations. This is a potentially hazardous area and it is important to maintain the highest standards of safety at all times. Access to the area controlled by an inner cordon should be restricted to the minimum numbers required for work to be undertaken safely and effectively. Personnel should only enter when they have received a full briefing and been allocated specific tasks.

At major incidents the fire brigade will retain a degree of responsibility for the safety of all personnel in the inner cordon. Special provision should be considered for non-fire service personnel (e.g. doctors) who need to operate in the area. These might include safety equipment and clothing, safety briefing, a record of their presence and agreed evacuation signals.

Outer cordons prevent access by the public into an area used by the fire service, and other services, for support activities. Outer cordons will usually be controlled by the Police.

Marshalling areas will usually be located within the outer cordon area if one is established.

1.2.4 Briefing & Information

Effective communication is of critical importance at all incidents. Information has to be relayed accurately from the Incident Commander to the crews carrying out the work and vice-versa so that the crews are aware of the tactics being employed and the Incident Commander is aware of what is happening on the incident ground. The Incident Commander also has a responsibility to relay messages and information back to Brigade Control to give more senior officers an accurate picture of the nature and demands of the incident.

The effective briefing of crews is essential. This may commence en route to an incident and will be supplemented on arrival, following an assessment of the incident and its associated risks by the Incident Commander. Following the initial assessment, crews will be briefed as to the tasks to be undertaken and the hazards and risks they will face. The extent of the briefing will depend largely on the nature and extent of the incident; the pre-briefing for small fires that crews regularly deal with is likely to be relatively straightforward. On the other hand, at incidents where crews have little experience and where there is a high risk factor, the briefing will need to be more comprehensive.

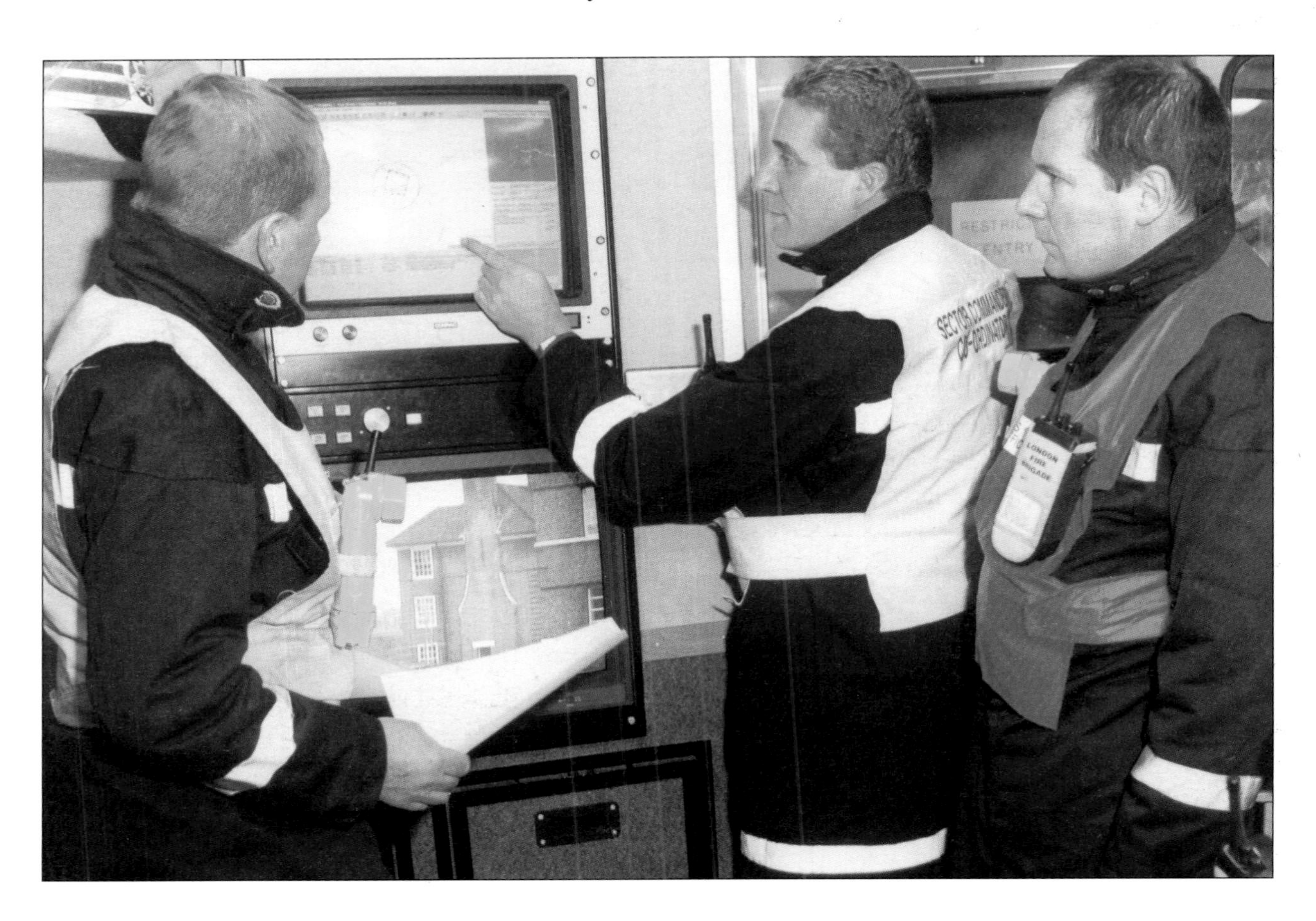

1.2.5 Managing Crews on the Incident Ground

On arrival crews should be kept together and, as far as possible, work as a team. An Incident Commander should be alert to the possibility that, due to enthusiasm, point of arrival on the incident ground, or for other reasons, crews can be tempted to self-deploy. This is bad practice, reduces accountability and robs the Incident Commander of resources which may be urgently required for other tasks. Brigades should adopt procedures to prevent this occurring.

Once crews have been briefed they must follow the brief and work safely. This will include wearing the appropriate personal protective equipment and ensuring that access and egress is properly secured. The Incident Commander will need to maintain a position where, so far as is practicable, he/she can observe proceedings. Where the risk level requires it, the appointment of one or more safety officers should be considered.

Once crews are at work they will require support. This means having the necessary resources available (e.g. BA cylinders and servicing facilities) and to ensure that their welfare needs are addressed. Care must be taken to give crews sufficient rest, relief and refreshment. The frequency of reliefs will depend upon the demands of the incident and brigade practice. A recognised problem exists at protracted rescues where personal commitment to the victims is high. Under these circumstances the crews' level of fatigue needs to be measured against their continued desire to work. A balance must be found between safe operations and crew morale.

The potential for post-incident stress is increasingly recognised and officers should be trained to identify individuals who may be susceptible and situations that may give rise to such problems. The need for support and counselling may need to begin on the incident ground.

1.2.6 Liaison

The Incident Commander must secure and maintain effective liaison with the other agencies which can contribute to resolving an incident. This will include liaison with other emergency services to co-ordinate activities effectively, and liaison with technical specialists whose specific knowledge may be critical in helping to respond to the risk. There is also a need to maintain effective liaison with the media, if in attendance, in order that appropriate accurate information is made available.

1.2.7 Post-Incident Considerations

The majority of fire service activities and interests centre around the emergency phase of an incident. However, there are issues which involve the fire service for well beyond the emergency phase. Examples include the following:

- Post-mortem enquiries and Coroner's hearings
- Fire Investigation
- Accident Investigation
- Public or judicial enquiries
- Litigation
- Financial costs to the brigade i.e. damaged equipment
- Criminal Investigation
- Incident Debriefing and Evaluation
- Fire Safety issues

The Incident Commander must, at the earliest possible moment, attempt to assess what the post-incident considerations might be. On the basis of this assessment, the following tasks might need to be undertaken:

Scene Preservation: As soon as it is identified that detailed examination of the scene might be required, efforts must be directed to preserve it from any further interference.

Recording and Logging: This might include a written log in the Command Unit during the incident or voice recording of critical messages. The early attendance and planned deployment of service photographic/video personnel can prove to be of great benefit in this area. The obtaining of security videos from on-site equipment can also often be of value in subsequent investigations. Action on this matter needs to be taken without delay as some systems will re-use the tapes.

Impounding Equipment: Where accidents or faults have occurred, any associated equipment should be preserved for later investigation. (Should major malfunction of Fire Service equipment occur, in addition to the normal required notification being carried out, Her Majesty's Fire Service Inspectorate should also be informed.)

Identification of Key Personnel: The names and location of witnesses to important events should be obtained and recorded for later interviews. It may be necessary or appropriate to commence interviewing during the incident.

1.2.8 Communications

The Incident Commander must establish effective arrangements for communications. Information is one of the most important assets on the incident ground; information must be gathered, orders issued and situation reports received. The needs of other agencies must be assessed and provided for.

The Incident Commander will need to:

(i) Establish communication links with Brigade control.

(ii) In accordance with the requirements of the Incident Command System, allocate fireground radios, assign channels and agree call signs.

(iii) Establish communications with other agencies. (This may employ communications equipment on agreed channels or simple direct verbal communication.)

(iv) Utilise local systems. Many new and complex buildings will have communication systems installed for emergency use.

Chapter 2 – Organisation on the Incident Ground

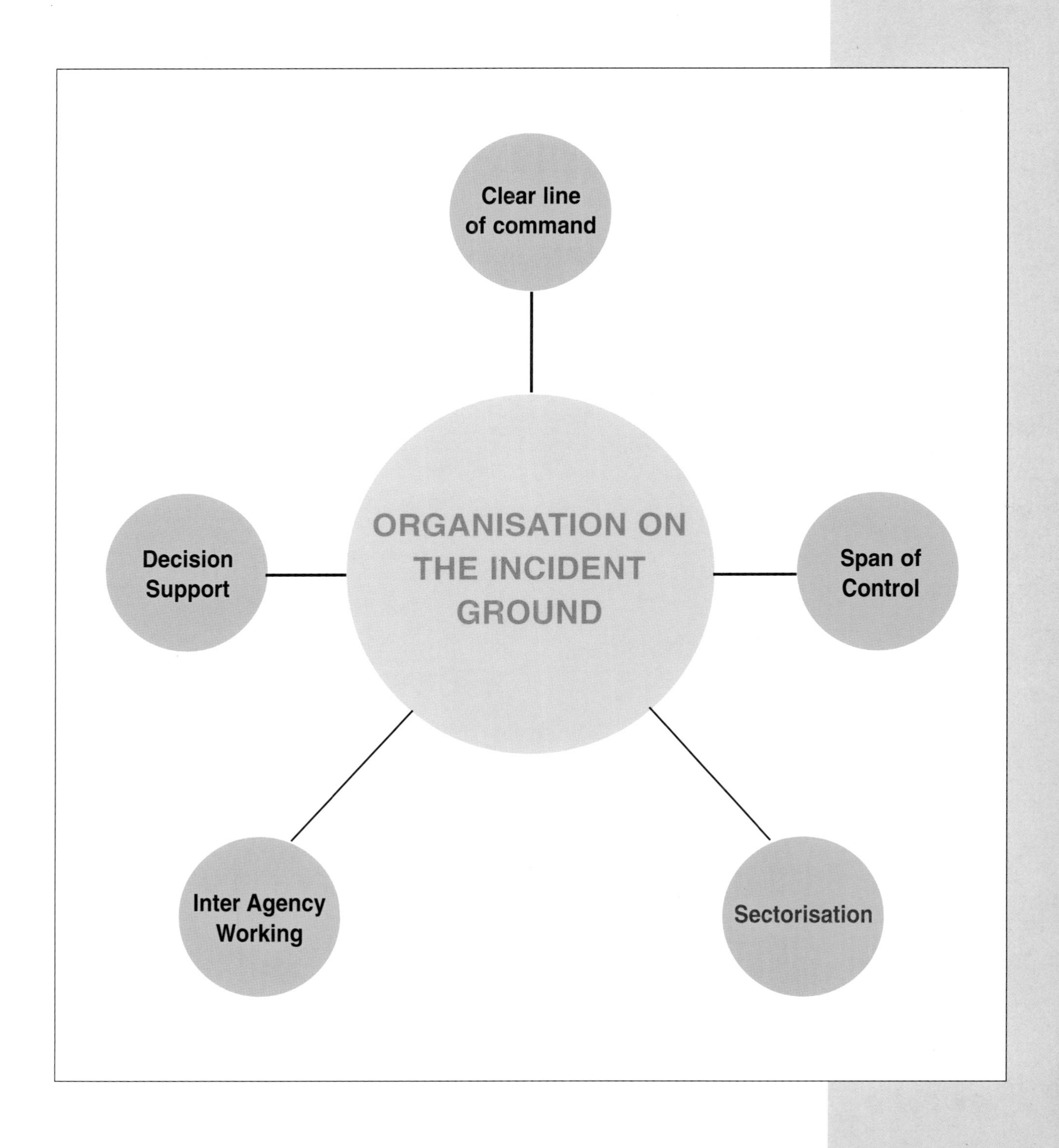

2.1 Incident Command Structure Explanation

The Incident Command System (ICS) is based on a framework which ensures manageable 'Spans of Control'. Other elements are built onto that framework. This provides the Incident Commander with the means to find a way through the complexity of the emergency situation and assists with the development of an effective and appropriate incident ground structure.

The concept of the 'Span of Control' is important to this basic structure. At a serious incident, during stressful and rapidly changing situations, an Incident Commander has to deal effectively with many people and a large amount of information. Therefore, the commander's span of control has to be limited. Sectorisation is central to the application of the principle of limiting spans of control and provides everyone on the incident ground with a clear line of reporting. The pattern of sectorisation must be both predictable and flexible [1].

Research has shown that, at most large incidents, Incident Commanders are not only making decisions about tactics, reinforcements, logistical problems etc., but also mentally building an organisation chart at the same time. The ICS provides a clear framework which expands from a one pump attendance to the largest incident that might ever occur and provides the Incident Commander with a ready to use organisational framework.

Terminology is important and it is necessary for everyone to use and understand a standard terminology. The system uses role titles, e.g., Sector One Commander, Incident Commander etc. This assists with good management and effective communications at incidents where only the fire brigade is involved and more so, for multi-agency incidents, where the 'Gold, Silver and Bronze (or 'Strategic, Tactical and Operational')' system is used. Inter-agency operations are also enhanced by all personnel appreciating where they feature in the complete incident organisation structure.

The other main elements of the standard framework are:

- A clearly defined and visible line of command.
- Management of the commander's span of control.
- Appropriately shared responsibility and authority, with clear definition and understanding of roles and responsibilities.
- Devolved information management and support for commanders; the 'Command Team'.

2.2 A Clear Line of Command

The system provides a framework for managing incidents based on a single, clearly defined, line of command which runs from the Incident Commander to every individual on the incident ground.

The line (or chain), of command is as described in the diagrams showing the framework of the ICS (see pages 19–26). The command framework is flexible enough to be adapted to incidents of any size and is based upon one essential element, i.e. that every unit on the incident ground, be it a crew or a sector has a single individual who is responsible for the effective management of that unit. **Every individual has responsibility for their own safety.**

2.3 Span of Control

The system requires that the direct lines of communication and areas of involvement of any officer be limited to enable the individual to deal effectively with those areas.

[1] For example, at a standard four-sided building, the front is one, rear three and sides two and four, in a clockwise pattern. High rises are sectorised by floor level. Unusual buildings or incident sites are sectorised by the Incident Commander upon arrival according to where resources are being deployed and what appears intuitively correct. Another important use is on the 'divided incident ground', where, due to geographical spread (forests, moorland etc) or separation due to spillage, toxic clouds or collapse, the Incident Commander is deprived of direct contact with the areas of operation in progress. At debriefs, the chosen pattern should be discussed to see what can be learned.

Span of Control

In this diagram an Incident Commander is responsible for 3 working crews at an incident and has detailed a firefighter to carry out a specific task which involves regular contact.

The span of control for this Incident Commander is 4. This being the number of lines of relatively constant communication which must be maintained.

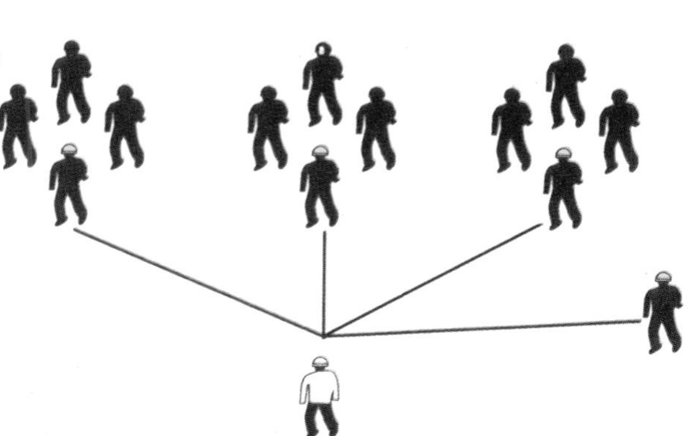

No individual should be responsible for so many aspects of the incident that it is difficult or impossible to give sufficient attention to each.

The system seeks to ensure that an appropriate 'Span of Control' is exercised at all times by the requirement for additional officers to be introduced into the chain of command when the demands on any individual's attention becomes excessive.

The span of control for tactical roles should be limited to **five** lines of direct communications, to ensure that commanders do not become overburdened. (The span of control for logistical roles, e.g. the Command Support Officer, may be wider).

In a rapidly developing or complex incident the span of control may need to be as small as 2–3 lines. In a stable situation, 6–7 lines may be acceptable.

At small incidents where the area of operations is easily manageable and there are no sectors, the Incident Commander may oversee all aspects of the incident directly.

2.4 Shared Responsibility and Authority – Roles in the Incident Command System

2.4.1 The Incident Commander

The Incident Commander will normally be the senior officer present at the incident according to each brigade's policy determining ranks and responsibilities at incidents.

On occasion a more senior officer may choose to attend an incident as an observer. That officer will, under the Fire Services Act, have overall responsibility for the incident, but need not assume the role of Incident Commander. The senior officer may opt to act as an advisor to the Incident Commander, reviewing tactical plans, assessing resource management and giving guidance as appropriate, but that officer will work directly with the Incident Commander and will not interrupt the chain of command. Naturally, a senior officer may choose to assume the role of Incident Commander should it be considered appropriate.

The Incident Commander is responsible for the overall management of the incident and will focus on command and control, deployment of resources, tactical planning, and the health and safety of crews At all incidents the Incident Commander will ensure that an individual is nominated as Command Support (see paragraph 2.4.4, page 13) and a contact point identified.

The Incident Commander, following an assessment of the incident, will allocate areas of responsibility to officers as necessary. These officers and specialist officers may be Sector Commanders. In the early stages of an incident the officers, of necessity, may be junior officers commanding appliances.

As the size and/or complexity of an incident increases, the demands on the attention of the

Incident Commander increase. While it is reasonable to assume an Incident Commander could manage a small house fire by dealing directly with all crews and individuals on scene, the same assumption cannot be made, for example, at a warehouse fire with crews working front and back, or when many crews are involved.

In order to manage the span of control effectively at larger incidents it will be necessary for the Incident Commander to delegate responsibility and devolve authority for some operations.

To achieve this the Incident Commander may choose to sectorise the incident. Sectors are created when the Incident Commander wishes to devolve responsibility for particular operations. Sectors can only be created when an individual of appropriate rank is available to assume responsibility for the operations within it. All crews within a sector should report directly to the Sector Commander.

In order to allow clear definitions of sector responsibility, sectors must have clearly defined boundaries. These may be topographic, usually the case for operational (as opposed to support) sectors, or functional, e.g. Water Sector, or Decontamination Sector.

2.4.2 The Sector Commander

To reduce the risk of confusion and to allow the proper assignment of tasks it is necessary for the boundaries of responsibility at an incident to be clearly defined.

This is best achieved by 'sectorisation'. A sector can be a physical area of the incident ground or an area of support operations. (see section 2.5)

It will be necessary for the Incident Commander to identify suitable areas of operations as sectors of responsibility and to designate each sector. A Sector Commander will be appointed for each sector. The Sector Commander will report to the Incident Commander, or where appropriate, to the Operations Commander.

It is imperative that a system of sector identification is used. Although the method of identification can vary (for example when dealing with complex premises or at incidents not easily conforming to the method normally employed), it is vital that whatever method is adopted by a brigade is used consistently at incidents and understood clearly by all personnel. Some examples of sectorisation are offered in section 2.6.

An officer assigned as a Sector Commander should assume the sector name as the incident ground radio call sign, e.g. 'Sector Two'. This identification of the sector names and their use as call signs can be extended to the functional support sectors. Examples would be 'Water' 'Decontamination' or 'Marshalling'.

At very large incidents the Incident Commander may appoint one or more Operations Commanders to take responsibility for a number of sectors.

2.4.3 The Operations Commander

The role of Operations Commander exists as a means of maintaining workable spans of control when the incident develops in size and complexity. If, for example, the incident has 4 operational sectors, some support sectors (e.g., water, decontamination, salvage, etc.) and there are also demands for the Incident Commander's time from press, specialist support, other services etc., the Incident Commander's span of control is likely to be at its upper limit. In this example, the operational sectors

can be condensed to one line of communication by using an Operations Commander in the way described in figures 6, 7 and 8.

The Operations Commander is a member of the Command Team. As such, the role is at the 'Silver' or Tactical level assisting the Incident Commander who is the fire service 'Silver'.

The Operations Commander's function is to co-ordinate the sectors and to exercise the Incident Commander's authority in that sphere. The Operations Commander must not become involved in support activities, e.g., management of support sectors, liaison with press or other matters etc; that should be dealt with by Command Support. The Operations Commander's role should be purely focused on supporting the Sector Commanders, co-ordinating their requests and requirements and monitoring safety and risk assessment.

It is important to note that if an incident does not demand the use of an Operations Commander because there are not enough sectors or activity is too low, then this extra tier is best omitted. There is no advantage in overstructuring an incident. Experience has shown that there is rarely a need for the Operations Commander rôle to come into play before a Principal Officer is in command of the incident.

At unusually large incidents, it may be necessary to use more than one Operations Commander to maintain span of control (see Figure 8). Such incidents are likely to be rare.

2.4.4 Command Support

Command Support should be introduced at all incidents to assist the Incident Commander in the management of the incident. At small to medium size incidents, the Incident Commander should nominate a junior officer or firefighter as Command Support who will operate from the designated contact point, which should be identified at every incident, usually by continuing to display flashing beacons.

Consideration should be given to identifying a contact point which is not involved directly in operations. An appliance not involved in pumping or an officer's car may be suitable for this purpose.

Command Support should initially provide, and maintain, radio communications between brigade control and the Incident Commander and will also have the following responsibilities:

- To act as first contact point for all attending appliances and officers and to maintain a physical record of resources in attendance at the incident.
- To operate the main-scheme radio link to the brigade control and to log all main-scheme radio communications.
- To assist the Incident Commander in liaison with other agencies.
- To direct attending appliances to an operational location or marshalling area as instructed by the Incident Commander and to record the status of all resources.

- To maintain a record of the outcome of the risk assessment and any review, as well as any operational decisions or actions taken as a result of it.

- To record sector identifications and officers duties as the assignments are made.

At a large incident, say above 5 pumps, a Command Unit is usually mobilised by brigades together with some form of support. An officer should head the Command Support Sector and be responsible for all areas of support to the Incident Commander. The position is described in all the following illustrations as **Command Support**.

The additional duties of Command Support at a large, escalating incident may well include:

- Arranging appliance positioning and parking to minimise congestion. This role may require close liaison with the Police or other agency in order to arrange for parked vehicles to be moved.

- Liaising with crews of specialist units to ensure optimum support to operational sectors.

- Arranging additional or specialist equipment and crews to Sector Commanders as required by the Incident Commander.

- Liaising with other agencies as necessary, including booking-in and supervision of their staff, managing the media etc.

- Briefing designated personnel.

- Arranging the reliefs of appliances and personnel.

The spans of control within Command Support should be continually monitored. Where appropriate a request for additional officers to assist should be made to the Incident Commander.

2.4.5 The Command Team

Incident Commanders cannot manage a complex and rapidly developing incident alone; effective and structured support is essential to successful operations.

A Command Team comprises the Incident Commander and whichever officers or staff are supporting that role. At the simplest level, this is the Incident Commander in charge of a one pump attendance, with Command Support often being the driver who is operating the radio. At a more complex level, the Command Team includes the expanded command support function.[1] Sector Commanders, although located in the operational sectors and widely dispersed, are still members of the Command Team.

2.5 Sectorisation of Incidents

Sectorisation should be considered when the demands on an incident make it imperative that responsibility and authority is delegated in order to ensure appropriate command and safety monitoring of all activities. As stated in paragraph 2.4.2, methods of sector identification can vary provided the method used is consistent and clearly understood by all personnel. The creation of sectors will only be done on instructions of the Incident Commander who will choose a sectorisation method appropriate to the demands of the incident.

Even if it is possible for the Incident Commander to oversee all operations, the need to sectorise will arise if there is so much going on that the Incident Commander risks being distracted and unable to give sufficient attention to each task. This would indicate that the Incident Commander's span of control is too great. If an Incident Commander's span of control is greater than about 5 lines of direct communication at a working incident, it is possible that performance will be adversely affected.

Frequently, operations take place in more than one location during an incident, for example at the front and rear of a building. In such cases the Incident

[1] Brigades will take different approaches to which roles and functions form part of the command team. This is natural as the aim is to integrate communications and decision making as seamlessly as possible between the Incident Commander and personnel engaged on tasks. Some of the command support functions may take place at location remote from the incident, particularly at major and multi-agency incidents.

Commander's span of control may only be 2 or 3 (to crew commanders). For instance, at a typical semi-detached house fire the Incident Commander has the ability to monitor tasks at front and back simply by moving too and fro; there is unlikely to be a need to sectorise. However, if the house is mid terrace and there is no quick access from front to back, then despite the small span of control, it is unlikely that the Incident Commander will be able to adequately manage operations and supervise safety at front and back. In this case the most appropriate response would be for the Incident Commander to retain command of the front of the building and any support activities, but to nominate a Sector Commander and assign all operations at the rear of the building to that sector.

It is important to note that where a Sector Commander has been appointed for the rear of a building this does not necessarily mean that a Sector Commander has also to be created for the front of a building if the Incident Commander is satisfied that he/she can retain a satisfactory level of command. It is quite acceptable for an Incident Commander to retain command of the majority of an incident in such circumstances.

Similarly, at a small RTA there will probably be no need to sectorise, but if crews are assigned to a car which has rolled 30 yards down an embankment while the main scene of operations is on the roadway, it may be necessary to assign that car as a separate sector.

The principle is that sectorising is driven by the need to delegate responsibility and authority in order to ensure appropriate command and safety monitoring of all activities. Except in exceptional circumstances, sectorisation should follow the preferred model. It is not necessary to sectorise just to make the incident conform with the sectorisation model.

2.5.1 Location of Sector Commander

It must be emphasised that Sector Commanders should stay in their sector. Sector Commanders pro-

vide direct and visible leadership at each sector and need to remain directly accessible to the crew commanders for whom they are responsible. Deviations from proper procedure would include Sector Commanders visiting the Command Unit, or touring the incident ground attempting to supervise operations in another sector whilst neglecting their own.

In the rare cases where it is essential that a Crew or Sector Commander leaves their post, at the direction of the Incident Commander, for a meeting, briefing or another purpose, he/she must be replaced to maintain continuity of command, supervision, safety, etc.

Support, or non-operational sectors, e.g. water, decontamination etc., are designated as the Incident Commander sees fit and may be grouped according to availability of officers and resources to suit the need.

One of the principal purposes of the ICS is to provide clear command and limited spans of control; to enable this to happen the proper line of command should be observed: a Sector Commander of an operational sector reports only to the Incident Commander, (or the Operations Commander if one is in place). Commanders of support sectors report to Command Support.

2.5.2 Exposures

A sector includes not only the face of the building in the sector, but also the exposure risk at the other side of the street, or the exposed parts of an adjoining building. The Sector Commander is responsible accordingly. Deployment of crews and requests for assistance should be made taking proper account of this.

2.5.3 Assuming and Handing-over Command of Sectors

When command of an incident changes it must be done formally. In the same way there must be a proper recognition of the appointment of a Sector Commander. The Sector Commander must be formally briefed on his duties by the Incident Commander, and on the status of operations in progress by the outgoing Sector Commander when taking over a sector.

The following diagrams are examples of how the ICS structure may be applied to incidents, and how the command structure expands to match the demands of an escalating incident.

The layout is not intended to be prescriptive, but certain features are considered 'standard'. For example, operational sectors are generally numbered, not named. All operational sectors report direct to the Incident Commander or, exceptionally, to the Operations Commander if one is in place.

All support sectors must report to the Incident Commander via the command support function. This is important to preserve spans of control. At more serious incidents, it is likely that the command support function will be headed by an officer of some seniority.

Although the diagrams give examples of five, eight, fifteen etc., pump incidents, this is illustrative only and the range is, of course, variable and dependent on the requirements of the incident.

Services and organisations listed to the right of the command support function, e.g., 'Police', 'Press' etc., are examples only and the list is far from exclusive. It may include any or all of the agencies who are stakeholders in the incident.

The larger number of crew members in the vicinity of Command Support that appear in the diagrams after a command unit is in place, represent those allocated as 'runners', radio operators etc. Individual brigades will have different ways of managing this requirement.

2.6 Examples of Sectorisation

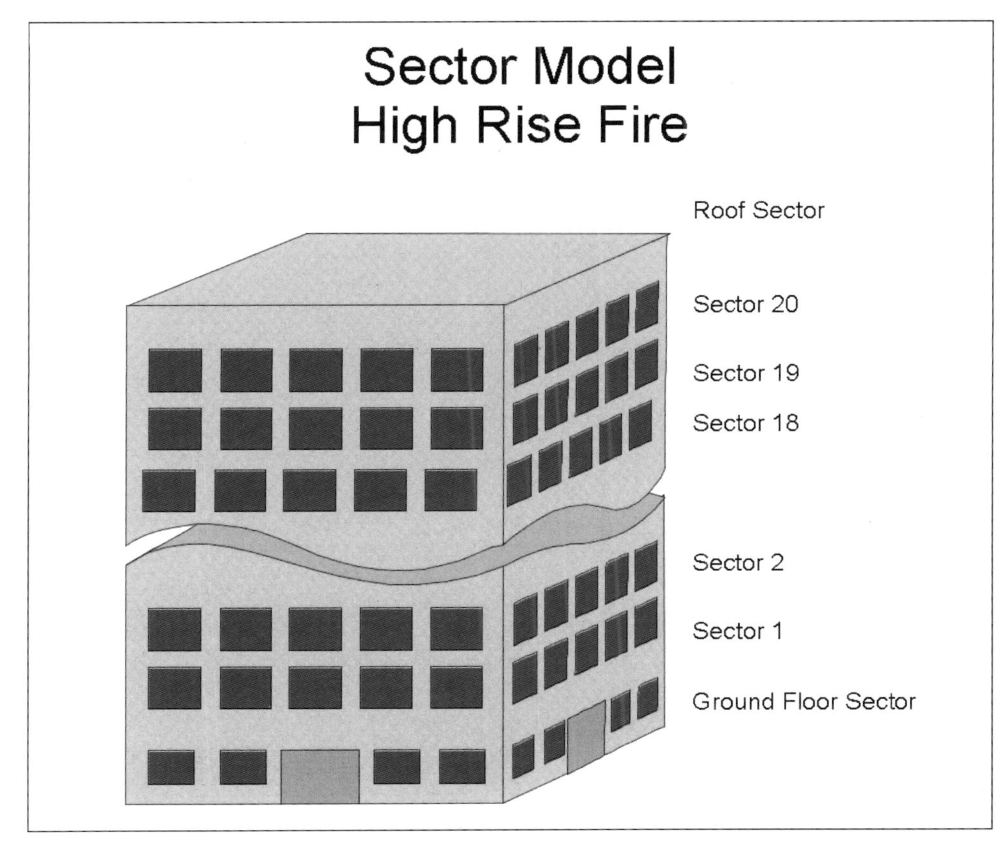

Sectorisation using floors in a building

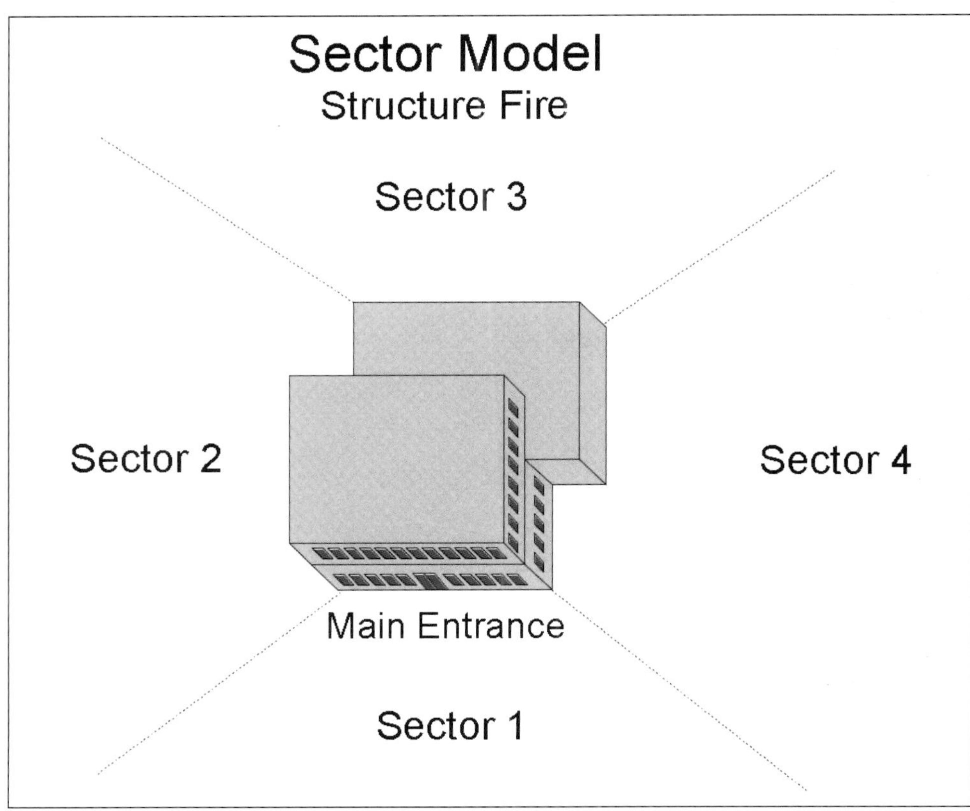

Sectorisation around a building

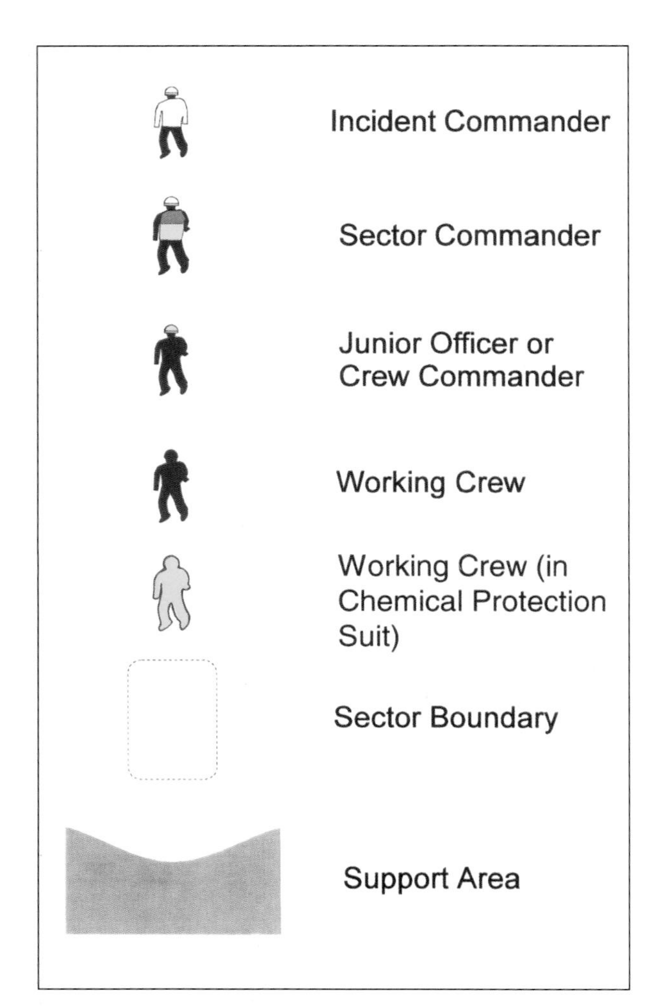

Key for the following diagrams

Figure 1

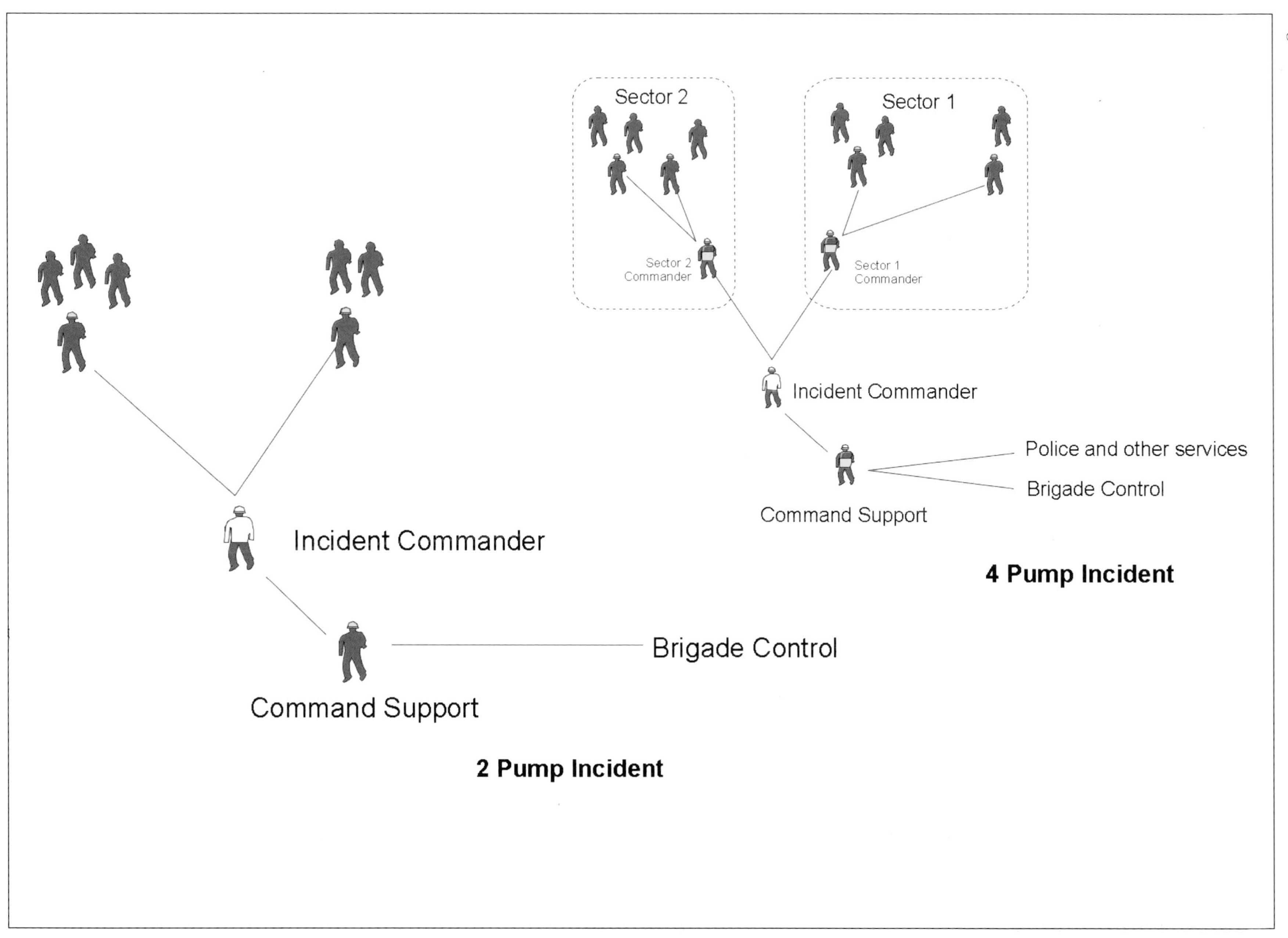

Sector 2
Sector 1
Sector 2 Commander
Sector 1 Commander
Incident Commander
Police and other services
Brigade Control
Command Support
4 Pump Incident
Incident Commander
Brigade Control
Command Support
2 Pump Incident

Figure 2

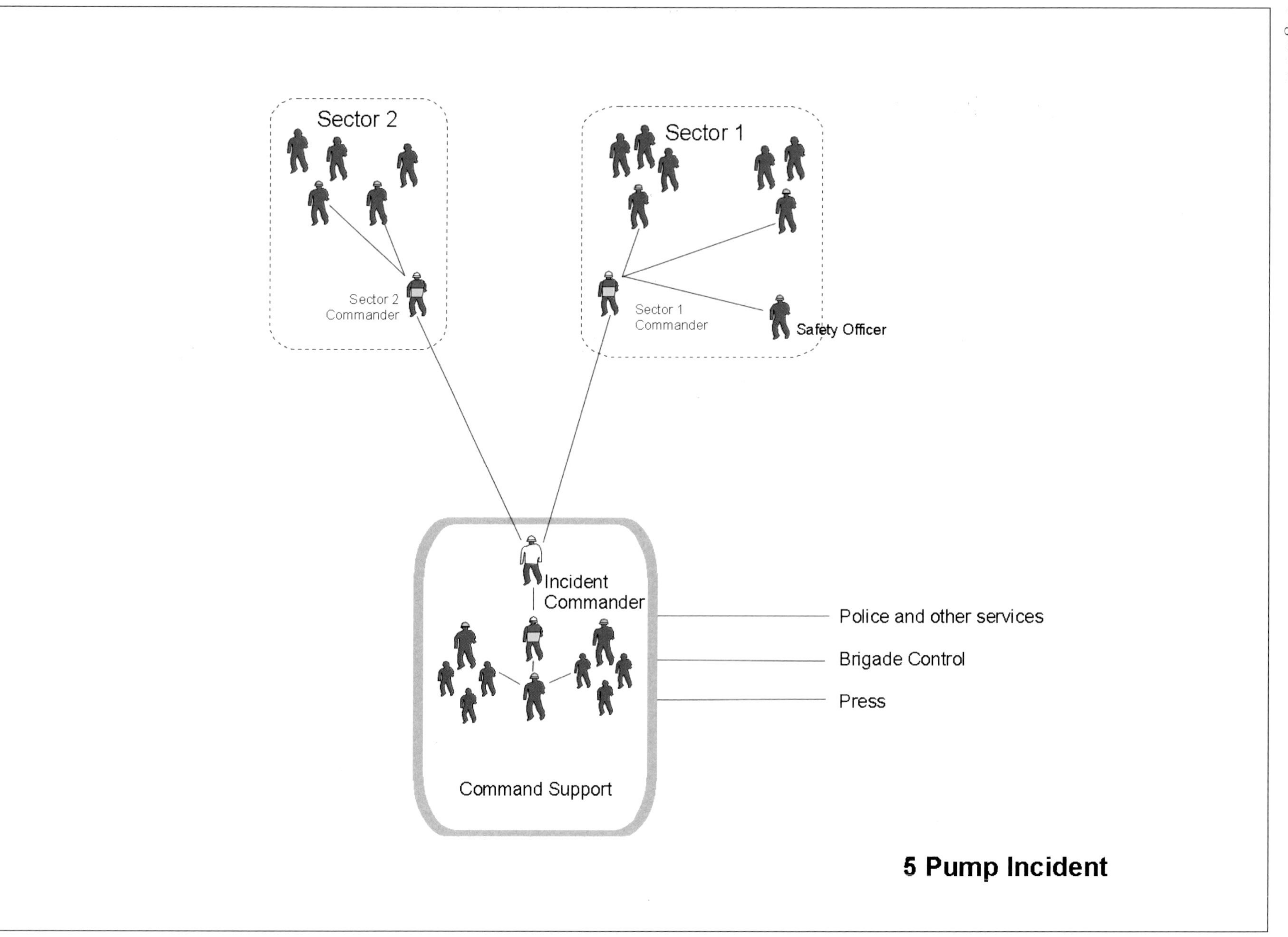

Sector 2
Sector 2 Commander
Sector 1
Sector 1 Commander
Safety Officer
Incident Commander
Police and other services
Brigade Control
Press
Command Support
5 Pump Incident

Figure 3

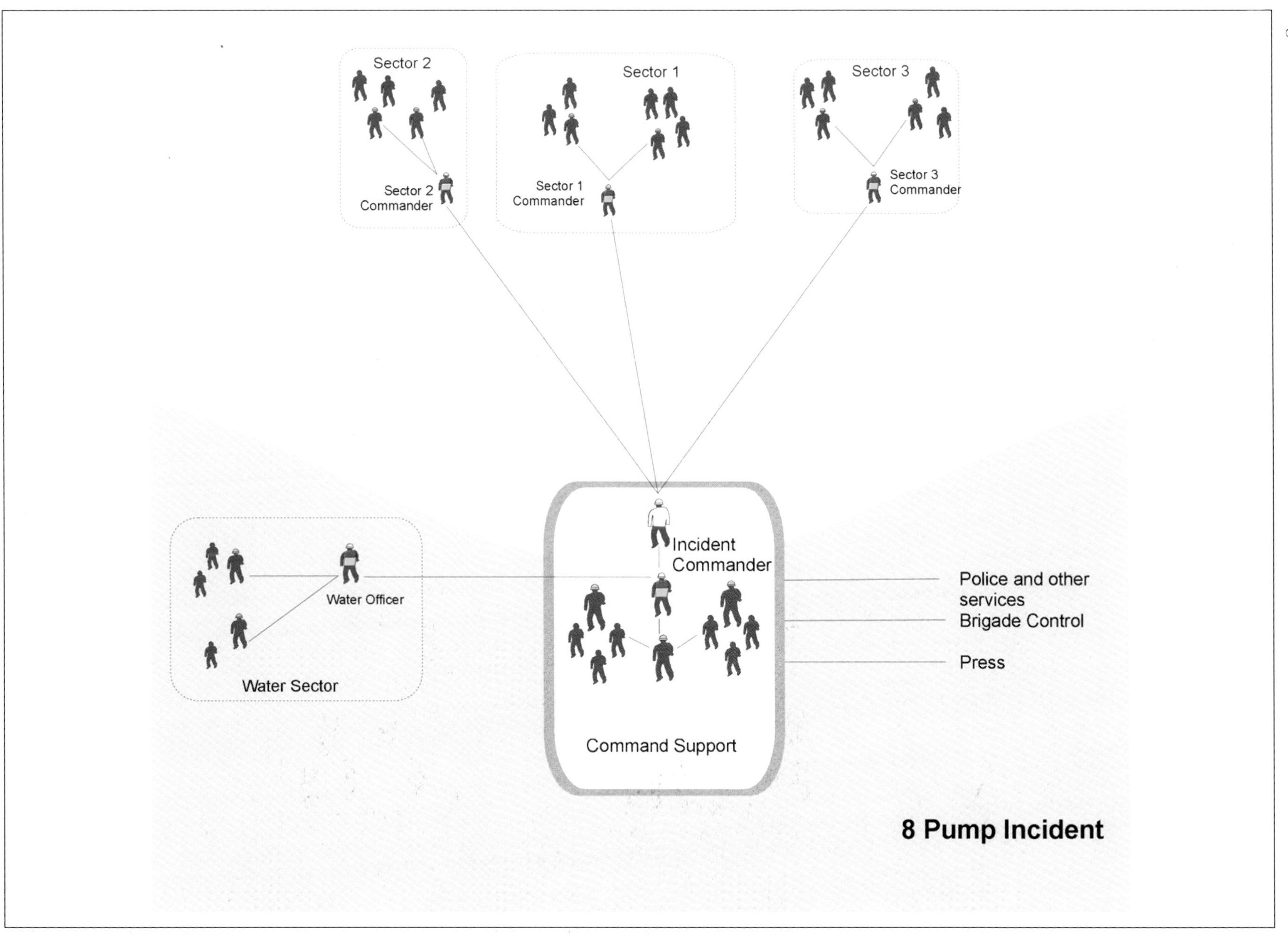

Sector 2
Sector 2 Commander
Sector 1
Sector 1 Commander
Sector 3
Sector 3 Commander
Incident Commander
Water Officer
Water Sector
Police and other services
Brigade Control
Press
Command Support
8 Pump Incident

Figure 4

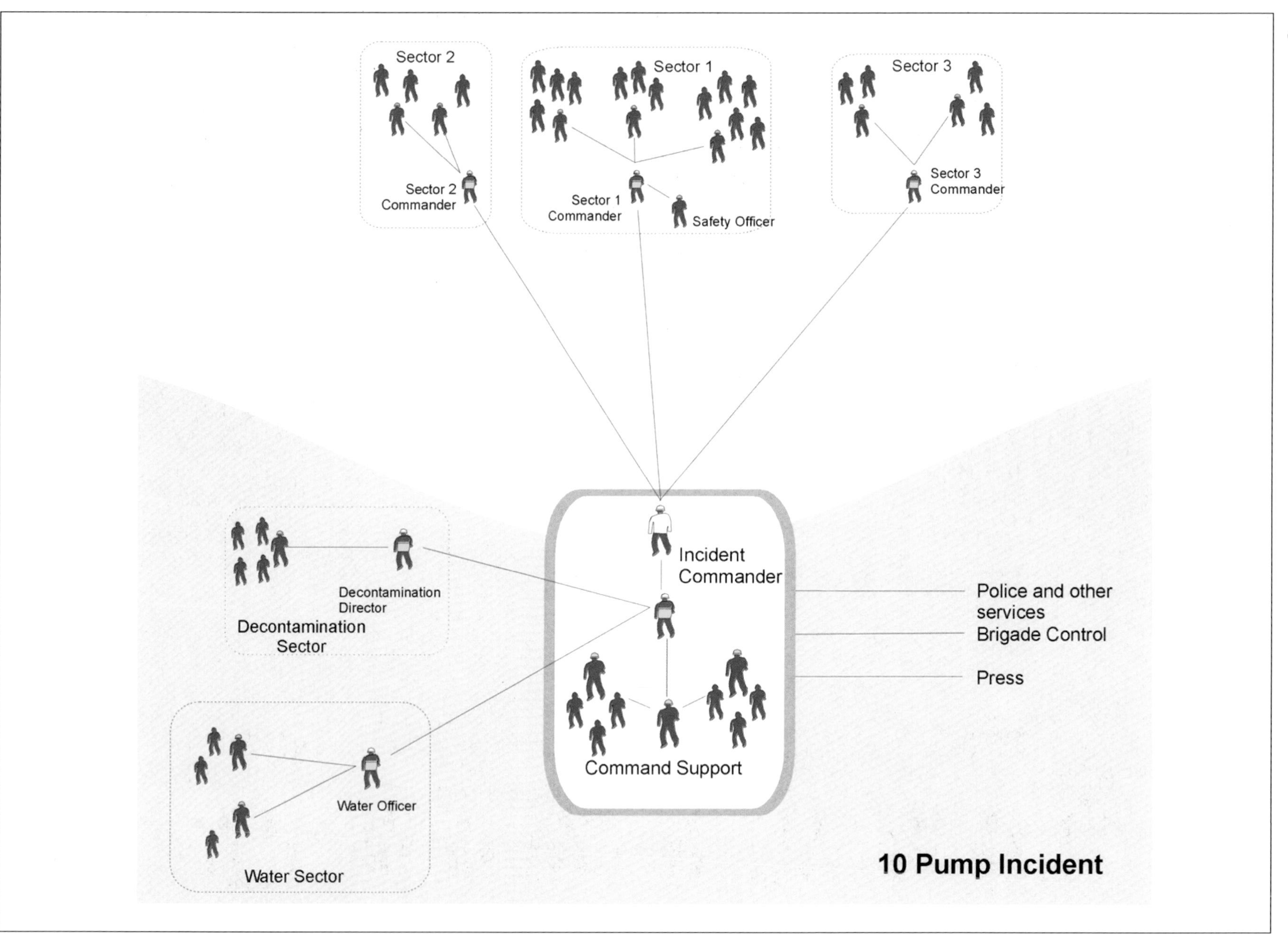
Sector 2
Sector 2 Commander
Sector 1
Sector 1 Commander
Safety Officer
Sector 3
Sector 3 Commander
Incident Commander
Command Support
Police and other services
Brigade Control
Press
Decontamination Director
Decontamination Sector
Water Officer
Water Sector
10 Pump Incident

Figure 5

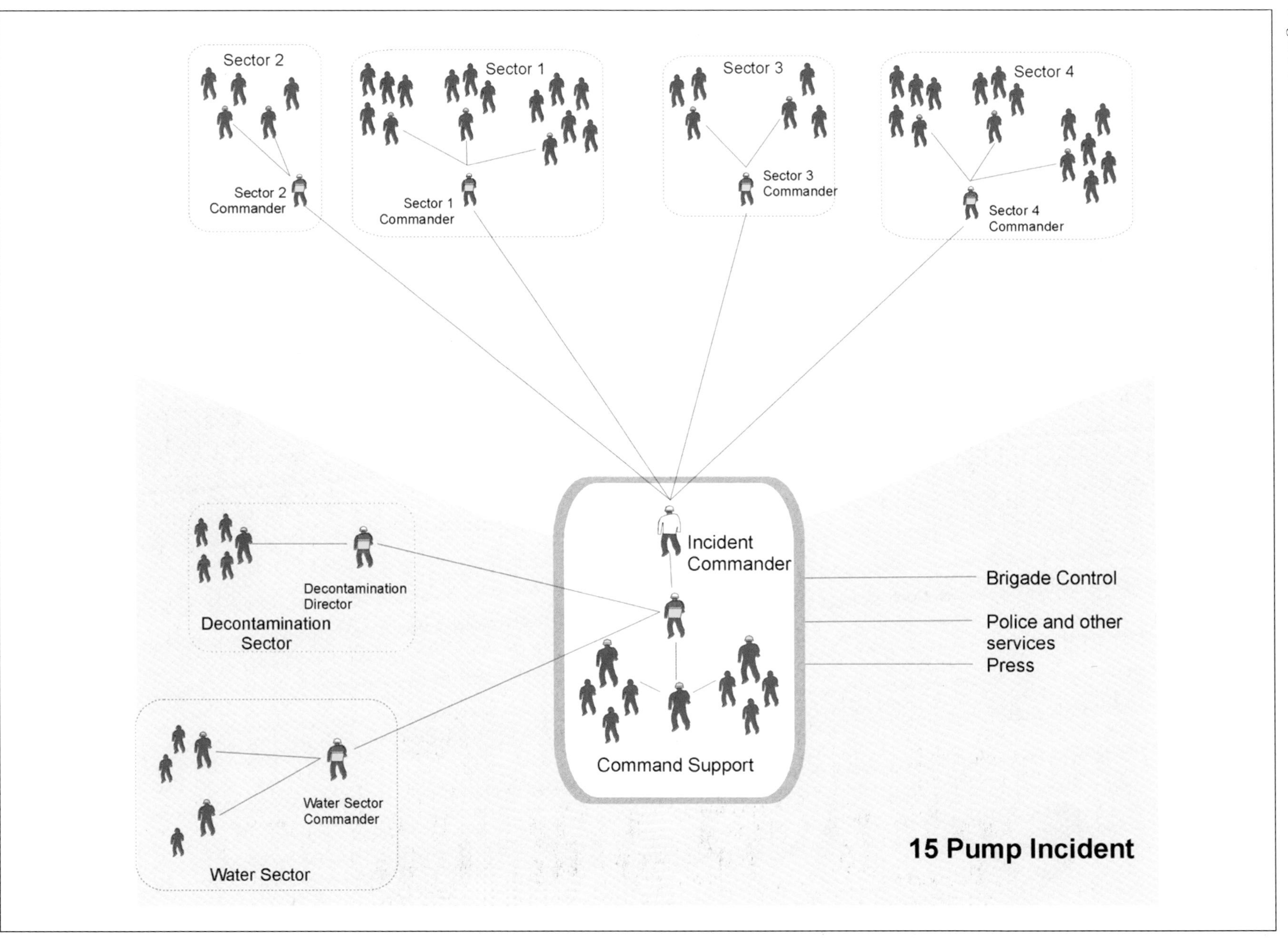
Sector 2
Sector 2 Commander
Sector 1
Sector 1 Commander
Sector 3
Sector 3 Commander
Sector 4
Sector 4 Commander
Incident Commander
Brigade Control
Police and other services
Press
Decontamination Director
Decontamination Sector
Command Support
Water Sector Commander
Water Sector
15 Pump Incident

Figure 6

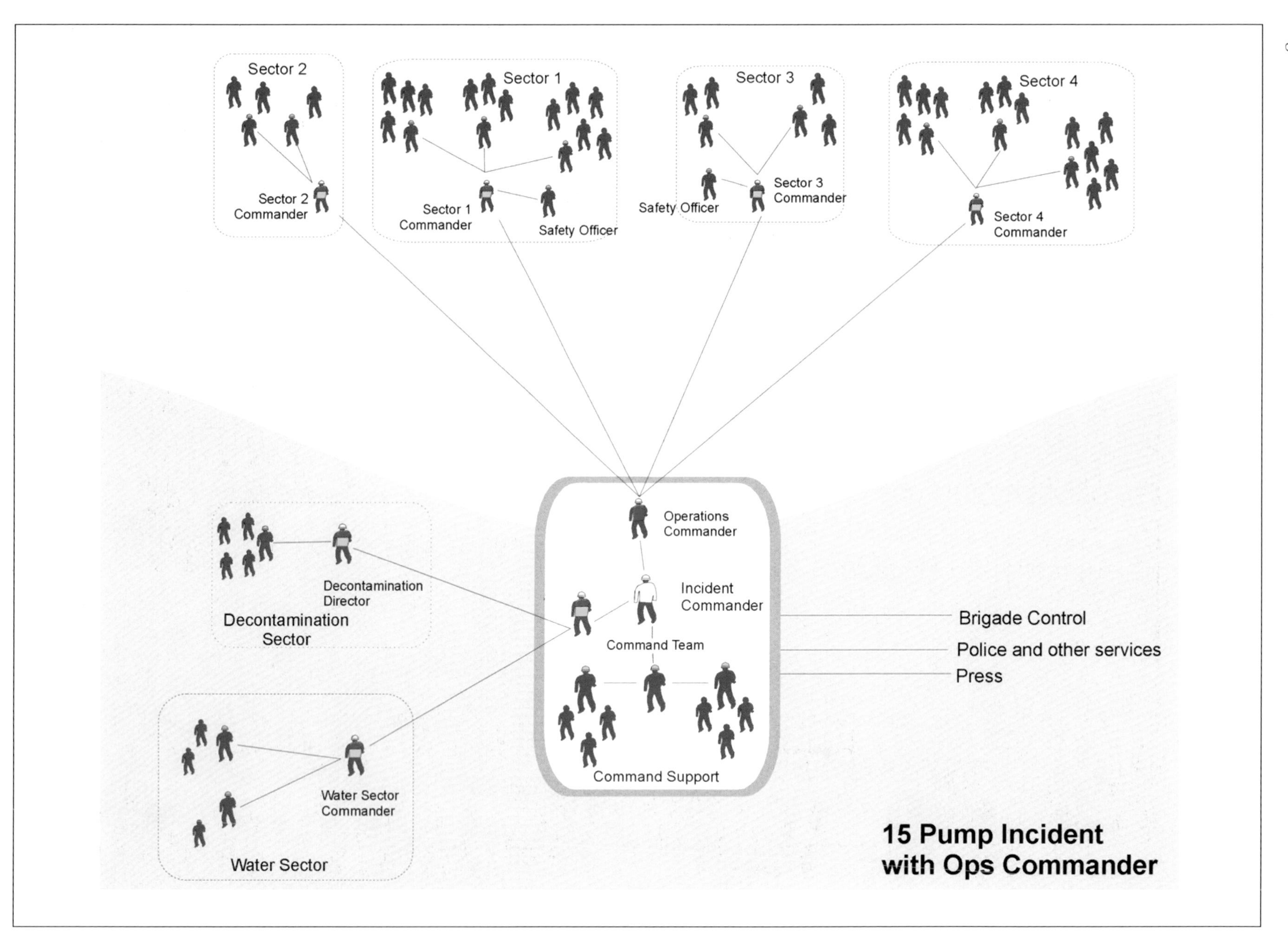

Sector 2
Sector 2 Commander
Sector 1
Sector 1 Commander
Safety Officer
Sector 3
Safety Officer
Sector 3 Commander
Sector 4
Sector 4 Commander
Operations Commander
Incident Commander
Command Team
Command Support
Decontamination Director
Decontamination Sector
Water Sector Commander
Water Sector
Brigade Control
Police and other services
Press
15 Pump Incident
with Ops Commander

Figure 7

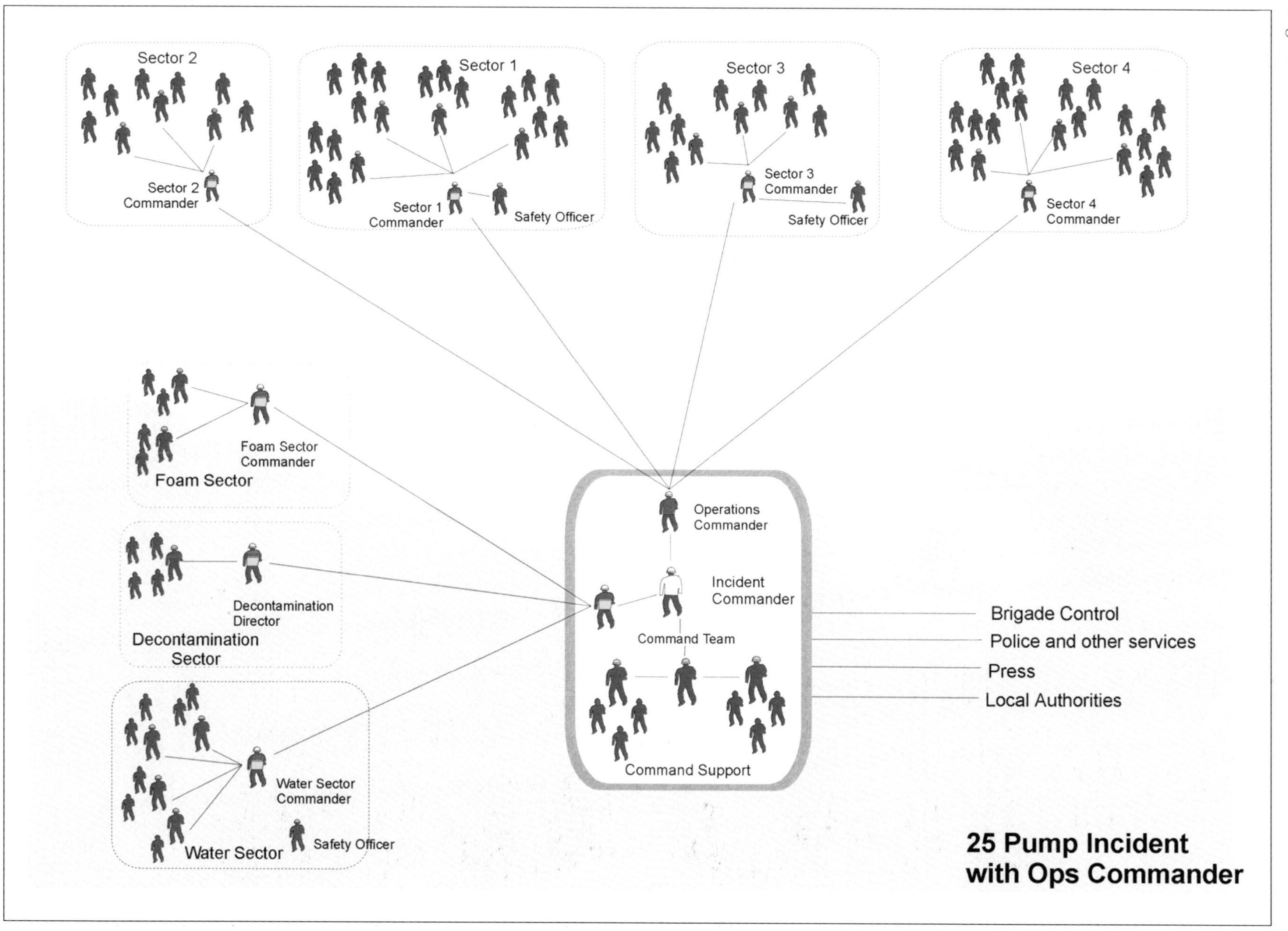

Sector 2
Sector 2 Commander
Sector 1
Sector 1 Commander
Safety Officer
Sector 3
Sector 3 Commander
Safety Officer
Sector 4
Sector 4 Commander
Foam Sector Commander
Foam Sector
Decontamination Director
Decontamination Sector
Water Sector Commander
Water Sector
Safety Officer
Operations Commander
Incident Commander
Command Team
Command Support
Brigade Control
Police and other services
Press
Local Authorities
25 Pump Incident
with Ops Commander

Figure 8

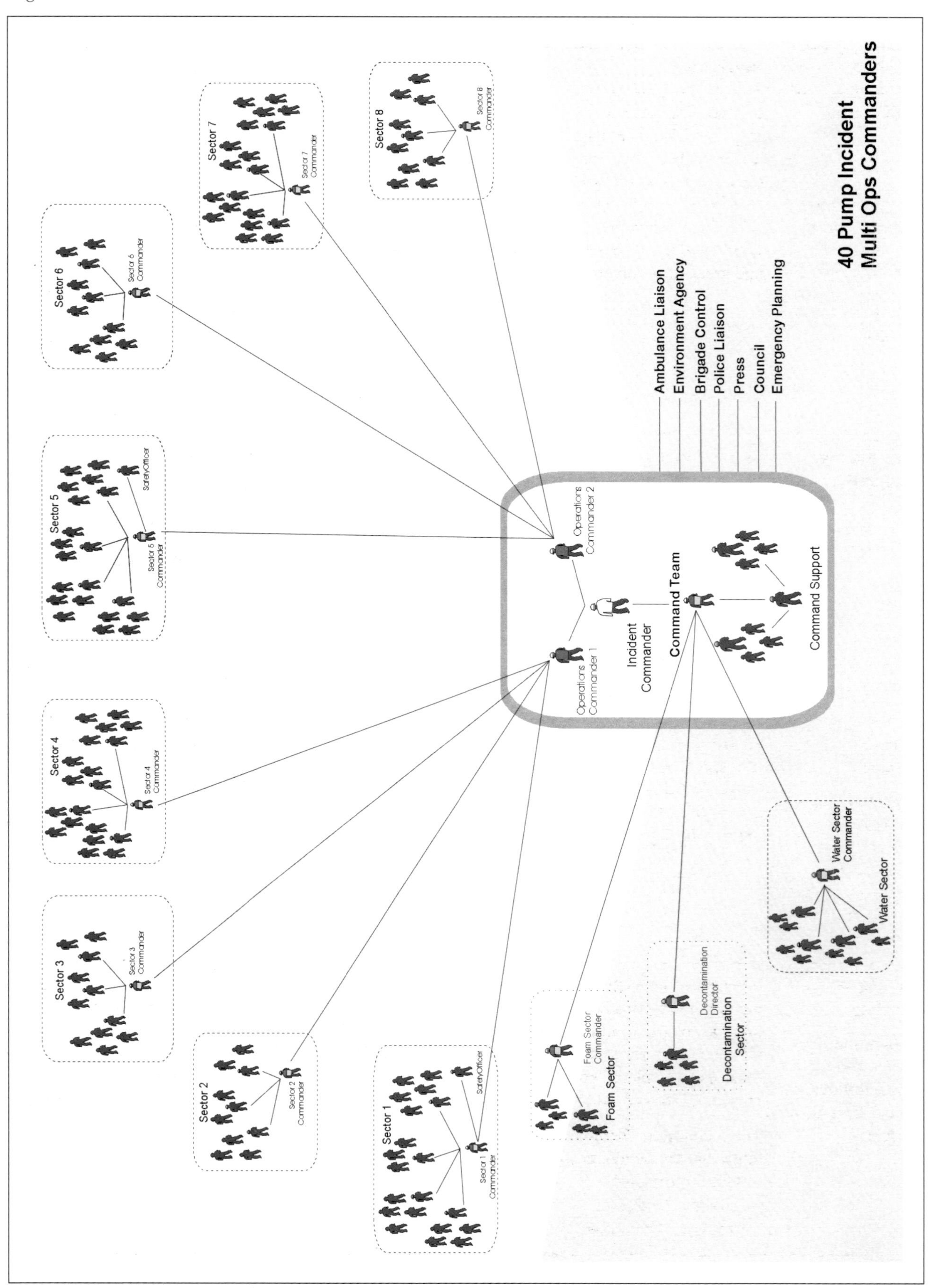

40 Pump Incident
Multi Ops Commanders
Ambulance Liaison
Environment Agency
Brigade Control
Police Liaison
Press
Council
Emergency Planning
Operations Commander 2
Operations Commander 1
Incident Commander
Command Team
Command Support
Sector 1
Sector 1 Commander
Safety Officer
Sector 2
Sector 2 Commander
Sector 3
Sector 3 Commander
Sector 4
Sector 4 Commander
Sector 5
Sector 5 Commander
Safety Officer
Sector 6
Sector 6 Commander
Sector 7
Sector 7 Commander
Sector 8
Sector 8 Commander
Foam Sector Commander
Foam Sector
Decontamination Director
Decontamination Sector
Water Sector Commander
Water Sector

ORGANISING THE INCIDENT: SUMMARY

Principle	Rationale	Comments
Sectorisation	It is necessary to clearly identify boundaries of responsibility. These areas or functions are known as SECTORS. A sector can be a physical area of the incident ground or a support function.	See para 2.5
Span of control	Effective management of incidents requires that the Incident Commander's direct lines of communication and areas of involvement must be limited.	Rapidly developing incident: 2–3 direct lines of communication. Stable situation at an incident: 6–7 lines (max). Ref: Ideally 4–5 lines of direct communication. National Fire Protection Association 1561 (1995)
Clear lines of command	Incident management roles should be clearly defined. Personnel allocated roles should be clearly identifiable. There must be a clear understanding of specialist command roles and responsibilities. Attendance of officers by rank may be linked to incident size and predetermined by a Brigade's mobilising protocols. It is not, therefore, possible to be prescriptive about rank. At some incidents, particularly in the initial stages, it may be necessary to have junior officers taking responsibility for large or complex command situations until more senior ranks arrive.	Principal Roles : Incident Commander Sector Commander Crew Commander
Decision Support Command Team; optimum command performance is dependent upon close support for the commander.	Further recognising the principle of span of control, the Incident Commander's responsibilities are shared by the Command team. Ideally, support would be in the form of a team of individuals with a common experience and training background.	Research links critical incident outcomes to efficient team building. (See appendix 3)
Inter Agency Working Recognition of roles and responsibilities at joint service incidents.	To ensure a consistent and co-ordinated approach to incident management where mutual assistance between brigades or where inter service resources are required, it is necessary for commanders in all of the attending services to be clear about the functions and responsibilities of other agencies involved. The recognised Command levels at largescale, multi-agency incidents are: Gold – Strategic Silver – Tactical Bronze – Operational	The Fire Service Major Incident Emergency Procedures Manual. Dealing with Disaster (3rd. Edition) Local authority emergency plans Local joint emergency service plans.

Chapter
3

Chapter 3 – Dynamic Risk Assessment

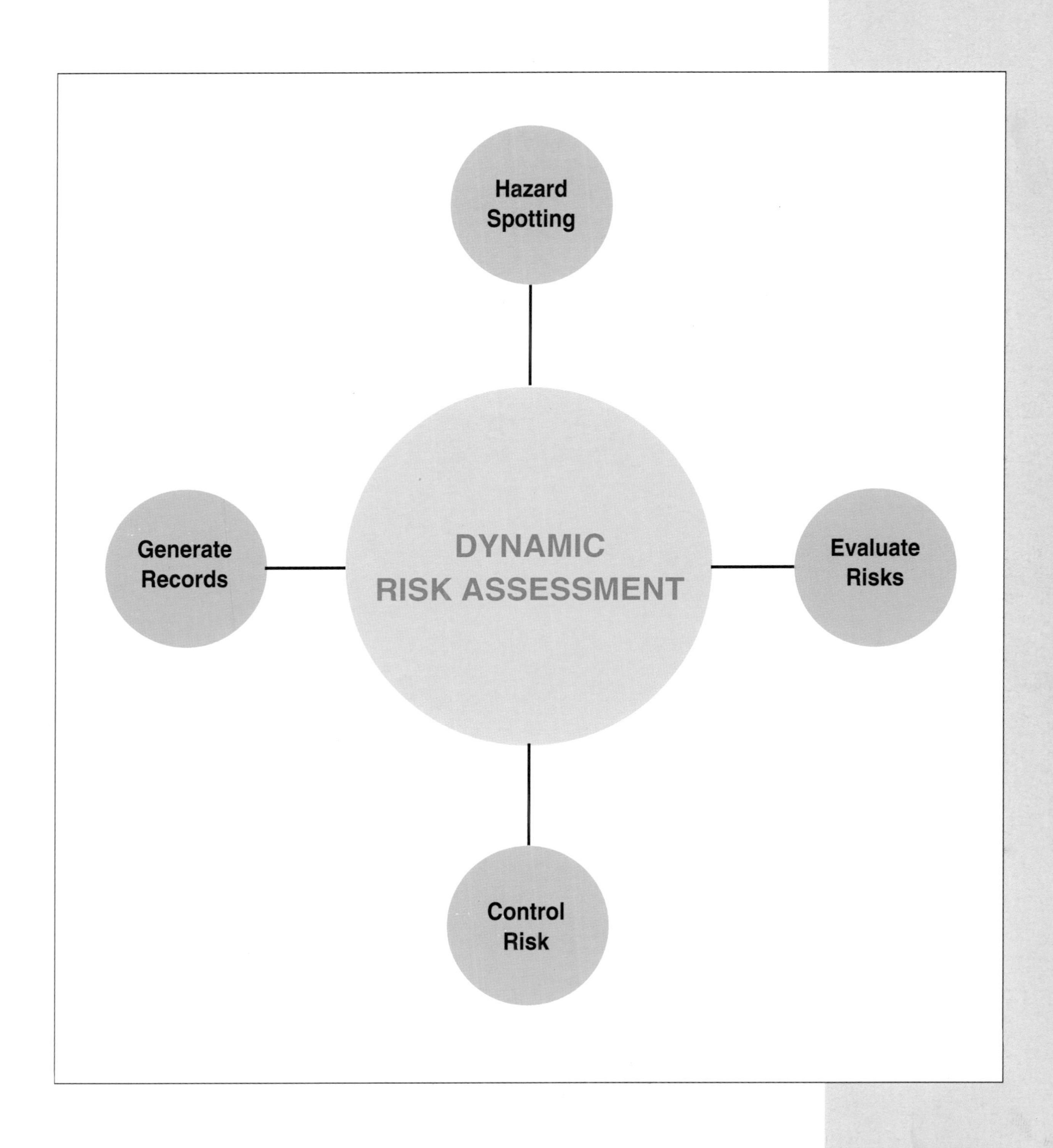

3.1 The Impact of Health & Safety on the Incident Ground

One of the principal and immediate considerations of the Incident Commander is the safety of all personnel. This must be established by assessing the hazards that are present and the possible risks to the health and safety of those at the scene and adopting appropriate, safe systems of work.

The philosophy of the fire service's approach to risk assessment can be summed up by the following:

- Firefighters will take some risk to save saveable lives.
- Firefighters will take a little risk to save saveable property.
- Firefighters will not take any risk at all to try to save lives or property that are already lost.

Regulation three of the Management of Health and Safety at Work Regulations 1992, requires that brigades carry out suitable and sufficient risk assessments of the risks to which operational personnel are exposed.

The term 'Dynamic Risk Assessment' is used to describe the continuing assessment of risk that is being carried out in a rapidly changing environment.

The key elements of any assessment of risk are:

- Identification of the hazards.
- Assessment of the risks associated with the hazards.
- Identification of who is at risk.
- The effective application of measures that control the risk.

When considering what control measures to apply, the Incident and Sector Commanders need to maintain a balance between the safety of personnel and the operational needs of the incident. For example, whereas it may be considered appropriate to commit personnel into a hazardous environment for the purposes of saving life, it may be that purely defensive tactics are employed in a similar situation where there is no threat to life.

The Incident Commander must ensure that safe practices are followed and that, so far as is reasonably practicable under the circumstances, risks are eliminated or, if not, reduced to the minimum commensurate with the needs of the task. However, because personnel may be working in sectors or smaller teams, everyone must be constantly aware of their own safety as well as that of their colleagues and others who may be affected by the incident or work activity.

3.2 Dynamic Risk Assessment

The term 'Dynamic Risk Assessment' is commonly used to describe a process of risk assessment being carried out in a changing environment, where what is being assessed is developing as the process itself is being undertaken. This is further complicated for the fire service commander in that, often, rescues have to be performed, exposures protected and stop jets placed before a complete appreciation of all material facts has been obtained.

Nevertheless it is essential that an effective risk assessment is carried out at any scene of operations. However, in the circumstances of emergency incidents, trials and experience have shown that it is impractical to expect the first arriving Incident Commander, in addition to the incident size-up and initial deployment and supervision of crews, to complete some kind of check-list or form.[1]

After action has been initiated on the basis of a 'Dynamic Risk Assessment', it is important that this is reviewed and confirmed as quickly as practicable, and further reviewed and confirmed at regular intervals.

Also, it is important that the outcome of a risk assessment is recorded, preferably in a way that is 'time stamped' for later retrieval and analysis, such

[1] For further information read Appendix 3

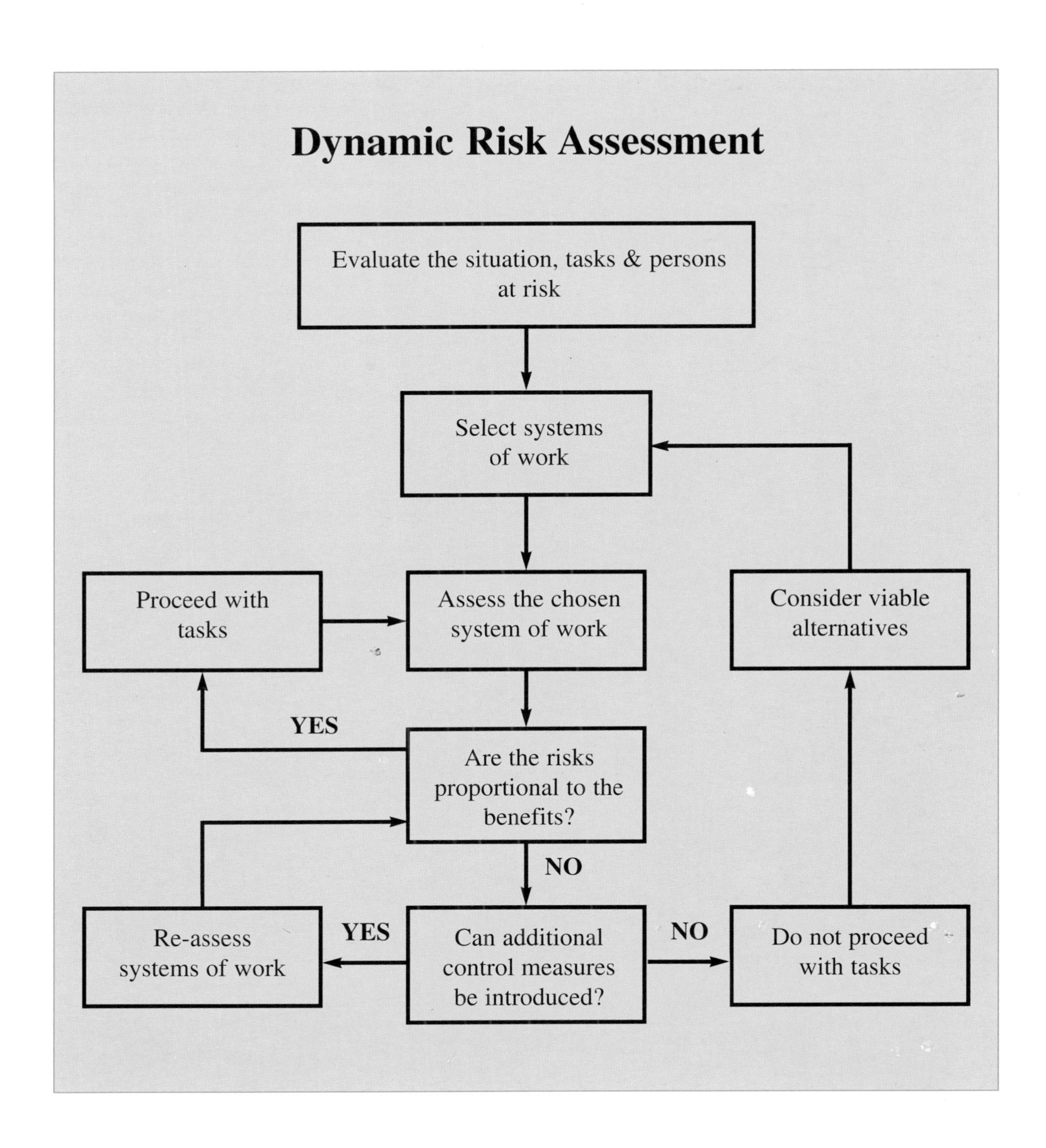

as would be achieved by transmission over the main scheme radio. (An example of how one Brigade has achieved this will be found in appendix 1).

Although the dynamic management of risk is continuous throughout the incident, the focus of operational activity will change as the incident evolves. It is, therefore, useful to consider the process during three separate stages of an incident:

- The Initial Stage.
- The Development Stage.
- The Closing Stage.

3.3 Initial Stage of Incident

There are 5 steps to the initial assessment of risk:

- Evaluate the situation, tasks and persons at risk.
- Select safe systems of work.
- Assess the chosen systems of work.
- Introduce additional control measures.
- Re-assess systems of work and additional control measures.

Step 1 Evaluate the situation, tasks and persons at risk

On the arrival of the initial attendance the Incident Commander will need to gather information, evaluate the situation and then apply professional judgement to decide the most appropriate course of action. Hazards must be identified and the risks to firefighters, the public and the environment considered.

In order to identify hazards the Incident Commander will initially need to consider:

- Operational intelligence information available from risk cards, fire safety plans etc.
- The nature of the tasks to be carried out.
- The hazards involved in carrying out the tasks.
- The risks involved to:
  firefighters,
  other emergency service personnel,
  the public and
  the environment.
- The resources that are available, e.g. experienced personnel, appliances and equipment, specialist advice.

Step 2 Select safe systems of work

The Incident Commander will then need to review the options available in terms of standard procedures. Incident commanders will need to consider the possible systems of work and choose the most appropriate for the situation.

The starting point for consideration must be procedures that have been agreed in pre-planning and training and that personnel available at the incident have sufficient competence to carry out the tasks safely.

Step 3 Assess the chosen systems of work

Once a course of action has been determined, be it offensive or defensive, Incident Commanders need to make a judgement as to whether or not the risks involved are proportional to the potential benefits of the outcome.

If YES proceed with the tasks after ensuring that:

- Goals, both individual and team are understood.
- Responsibilities have been clearly allocated.
- Safety measures and procedures are understood.

If NO continue as with step 4 below.

Step 4 Introduce additional control measures

Incident commanders will need to eliminate, or reduce, any remaining risks to an acceptable level, if possible, by introducing additional control measures, such as:

- Use of Personnel Protective Equipment e.g. safety glasses, safety harnesses.
- Use of BA.
- Use of specialist equipment e.g. HP, TL.
- Use of Safety Officer(s).

Step 5 Re-assess systems of work and additional control measures

Even when safe systems of work are in place, there may well be residual risks. Where these risks remain, the Incident Commander should consider if the benefit gained from carrying out the tasks against the possible consequences if the risks are realised:

- If the benefits outweigh the risks, proceed with the tasks.
- If the risks outweigh the benefit do NOT proceed with the tasks, but consider viable alternatives.

3.4 Development Stage of Incident

If an incident develops to the extent that sectors are designated, the Incident Commander will delegate the supervisory role to Sector Commanders. They will be responsible for the health and safety of all personnel within their sector.

Sector Commanders may feel that they can supervise safety within their own sectors. Alternatively, after consideration, the Sector Commander may feel it necessary to nominate a safety officer. This officer will be responsible to the Sector Commander.

As the incident develops changing circumstances may make the original course of action inappropriate, for example:

- Fire-fighting tactics may change from defensive to offensive.
- New hazards and their associated risks may arise e.g. the effects of fire on building stability.
- Existing hazards may present different risks.
- Personnel may become fatigued.

Both Incident and Sector Commanders, therefore, need to manage safety by constantly monitoring the situation and reviewing the effectiveness of existing control measures.

3.5 The Closing Stage of Incident

The three key activities involved in the closing stages of an incident are:

- Maintaining Control.
- Welfare.
- Incident debrief.

Maintaining Control

The process of task and hazard identification, assessment of risk, planning, organisation, control, monitoring and review of the preventive and protective measures must continue until the last appliance leaves the incident ground.

There are usually fewer reasons for accepting risks at this stage, because there are fewer benefits to be gained from the tasks being carried out. Incident and Sector Commanders should therefore have no hesitation in halting work in order to maintain safety.

As the urgency of the situation diminishes, the Incident Commander may wish to nominate an officer to gather information for the post incident review. Whenever possible, this officer should debrief crews before they leave the incident, whilst events are still fresh in their minds.

Details of all 'near misses' i.e. occurrences that could have caused injury but did not in this instance, must be recorded because experience has shown that there are many near misses for every accident that causes harm. If, therefore, we fail to eradicate the causes of a near miss, we will probably fail to prevent injury or damage in the near future.

Welfare

The welfare of personnel is an important consideration. It must be given particular attention by the command team at arduous incidents or incidents that require a rapid turnover of personnel. The physical condition of crews must be continually monitored by supervisors.

Welfare includes provision of rest and feeding which should, where possible, be outside the immediate incident area and always away from any risk of direct or indirect contamination.

Incident Debrief

Following an incident any significant information gained, or lessons learned, must be fed back into the policy and procedures of the brigade. Points to be covered may be in relation to existing operational intelligence information, personal protective equipment, the provision and use of equipment, other systems of work, instruction, training, and levels of safety supervision etc.

It is important to highlight any unconventional system or procedure used which was successful or made the working environment safer.

It is equally important to highlight all equipment, systems or procedures which did NOT work satisfactorily, or which made the working environment unsafe.

3.6 Summary of the Safety Function

- Identify safety concerns.
- Initiate corrective action.
- Maintain safe systems of work.
- Ensure all personnel are wearing appropriate personal protection equipment.
- Observe the environment.
- Monitor physical condition of personnel.
- Regularly review.
- Sector Commanders or nominated Safety Officers should update the Incident Commander of any changing circumstances.

DYNAMIC RISK ASSESSMENT: SUMMARY

Principle	Rationale	Comments
Identification of Incident Ground as a workplace.	The need for Safe systems of work applies on the Incident Ground as any other workplace. A clearly identifiable and communicated series of control measures are developed and implemented to reduce risks to an acceptable level.	HSG65 and BS8800 refer.
H.O. Risk Assessment guidance: The dynamic risk assessment.	All activities should be conducted so as to minimise the risks to operational personnel and the public. All tactical and operational deployments must be based on a risk assessment. Where a significant risk (e.g., a chemical hazard or a dangerous structure) is identified steps must be taken to manage that risk through the application of pre-planned standard operational procedures.	See para 3.2
Record of risk assessment: Recording and promulgation of the outcome of the Risk Assessment.	All personnel on the incident ground must be made aware of the outcome of the risk assessment in each sector and the overall incident, and be made aware of the standard operational procedures put in place to manage the risk. The outcome of the risk assessment and the application of the relevant standard operational procedure must be recorded.	See para 3.1

several other issues feature in most brigades' operational procedures. Each brigade will need to select the elements they require and add, adapt or develop them as necessary to meet local circumstances. What *must* be consistent, however is the overall framework.

COMMAND COMPETENCE AND TRAINING FOR COMPETENCE: SUMMARY

Principle	Rationale	Comments
Acquiring competence in command.	Officers should be trained in incident command competences.	Fire Service College progression courses. Brigade training systems computer aided training systems. EFSLB standards and draft standards. FSC 15/97 "A Competence Framework for the Fire Service"
Application of competence in command.	Ability to apply core competences should be regularly assessed.	EFSLB standards and draft standards. FSC 15/97 "A Competence Framework for the Fire Service"
Continuing development.	Performance can only be improved by taking account of lessons learned at incidents, exercises and simulation. Brigades should have an effective debriefing procedure which facilitates feedback into content / structure of training and operational procedures.	HSG65

Appendices

APPENDIX 1 – An Example of an Operational Risk Assessment Process

It is essential that an effective risk assessment is carried out at any scene of operations. It is completely impractical, however, to expect the first arriving Incident Commander to complete some kind of check-list or proforma. Historically, risk assessment on the incident ground was carried out during 'size-up' and initial deployment. What was lacking was consistency in methodology, the ability to demonstrate the process and a hard record of its outcome.

When considering the management of risk on the incident ground, it is essential that tactical co-ordination is of the highest order. The need for this increases considerably the more a commander delegates responsibility for tactical decision making.[1]

To ensure that all parts of the incident are managed effectively and safely an early and ongoing task for the commander is to decide and declare the tactical approach to the incident. Additionally, it is desirable that there is a contemporaneous record of the overall risk assessment. The following provides details of how this is applied in one brigade. (See chapter 3 'Dynamic Risk Assessment' in the main text.)

The Operational Risk Assessment Process in the West Yorkshire Fire Service: The Tactical Mode

A1.1 Incident Command System – The Tactical Mode

The Tactical Mode procedure assists the Incident Commander to manage an incident effectively without compromising the health and safety of personnel by:

1 An illustration of this problem would be where the Sector One Commander sees a situation where, despite a structure which is in danger of collapse, persons need rescuing. Therefore, crews are deployed in BA to perform that task: the sector is 'offensive', with crews committed into the building. At the same time in sector three, the Sector Commander sees a situation where the building is unstable: he/she would not and does not need to commit crews and therefore elects to adopt a defensive mode with external main jets and aerial monitors. This presents the hazard of external monitors bringing structure down around the crews deployed by sector one because the sectors were working against each other!

- Ensuring that firefighting operations being carried out by a single crew, or sector, do not have adverse effects on the safety or effectiveness of firefighters in other crews or sectors. (For example, it will ensure that BA wearers inside a building are not subjected to an aerial monitor being opened up above them, or to the impact of a large jet through a window from another sector without warning).

- Generating a record of the outcome of the dynamic risk assessment process conducted by the Incident Commander.

There are three **Tactical Modes**:

(i) **Offensive** – This mode may apply to a sector, or the entire incident.

This is usually applied where the operation is being dealt with internally, with the objective of carrying out search and rescue or fighting the fire before it involves the whole building or threatens its stability.

Offensive Mode is the normal mode of operation used at, for example, house fires and industrial premises to fight the fire, effect rescues, or close down plant, etc.

(ii) **Defensive** – This mode may apply to a sector or the entire incident.

This mode is generally employed for external operations and must be applied where committing firefighters internally would constitute an unnecessary risk to life. (For example: at a fire which has fully involved an evacuated large uncompartmented building, or in a building that is displaying signs of collapse).

In these circumstances the Incident Commander would adopt the **Defensive Mode**, fight the fire with external and aerial jets, and protect exposure risks and adjoining property.

(iii) **Transitional** – This mode may only apply to the whole of the incident and not to individual sectors alone.

This mode is used where the Incident Commander intends a shift in the mode of operations or where a combination of both Offensive and Defensive modes are in operation within different sectors at the same incident.

A 'Transitional Mode' would be adopted, for example, where:

- A building fire being fought externally with sectors in Defensive Mode has an annex that can be saved, safely, by using an Offensive Mode i.e., by fighting the fire inside the annex. Here there would be, say, three sectors in Defensive Mode and one in Offensive: the incident would be Transitional.

- A Defensive approach is being utilised only as an interim measure, until further resources arrive which will enable the Offensive Mode to be established and an attack on the fire to commence.

- An Offensive Mode is in effect but the building is in danger of becoming unstable, resources are being deployed in preparation to switch to Defensive Mode *if* this becomes necessary, e.g.: water lines being laid ready to supply aerial monitors.

- An offensive approach is in effect but circumstances dictate that an evacuation and withdrawal of equipment is necessary in order that a Defensive Mode can be utilised.

Further examples are included at the end of this Appendix.

A1.2 The Application of Tactical Mode

The Tactical Mode should be stated at all working incidents.

As the incident grows and the Incident Commander's span of control increases, it is essential that all personnel are aware of the tactics on the incident ground and the prevailing Tactical Mode.

All communications with Brigade Control will include a confirmation of the Tactical Mode for the information of oncoming appliances and officers.

A typical Informative Message might be 'Informative message from ADO Black at Green Street, Anytown, Factory premises, used for textile manufacturing, three floors, 20m × 20m. Ground and first floor well alight, three large jets in use, "WE ARE IN DEFENSIVE MODE" '.

A1.3 Adopting a Tactical Mode when Sectors are in use

When the incident has been divided into sectors, **the Incident Commander will retain responsibility** for the Tactical Mode at all times.

Sector Commanders must be made aware of any intervention by the Incident Commander to initiate change of the Tactical Mode. Sector Commanders may then implement the change effectively and ensure that personnel under their command are aware of the prevailing Tactical Mode.

There will be occasions when Sector Commanders wish to change the Tactical Mode in their sector. For example; they may detect signs of collapse or obtain information about some previously unknown danger. In such circumstances, they must take the necessary action for the safety of the crews and then advise the Incident Commander of the developments.

If, alternatively, the Sector Commander wishes to commit personnel internally in 'Offensive Mode' when the prevailing mode is 'Defensive', the permission of the Incident Commander must be sought and no change made until it is granted.

The Incident Commander will determine whether the Tactical Mode will change to Offensive on the whole incident ground, or just in that sector, making the mode Transitional. This decision will be based on an understanding of the status of operations in all other sectors.

It is essential that regular communication is maintained with the Incident Commander, who will assess the viability of such a request, having taken into account the possible effects of fire fighting operations in other areas.

A1.4 Responsibilities for determining Tactical Mode

A1.4.1 Incident Commander

To make an assessment of the incident and decide which Tactical Mode will be appropriate.

Advise Brigade Control which Tactical Mode is in operation at the incident and ensure confirmation is repeated at about twenty minute intervals up until the time that the 'stop' message is sent and at appropriately regular intervals thereafter.

The Incident Commander should consider and confirm the Tactical Mode on initial and all subsequent briefings to Crew and Sector Commanders.

A1.4.2 Sector Commanders

To continually monitor conditions and operational priorities in the sector and ensure that the prevailing Tactical Mode continues to be appropriate.

To immediately react to adverse changes, withdrawing personnel from risk areas without delay if necessary. To advise the Incident Commander of the change in conditions as soon as possible thereafter.

To consider if and when it is appropriate to change Tactical Mode and to **seek the permission of the Incident Commander to do so**. (Before giving this permission, the Incident Commander will determine the status of all other operational sectors to ensure that nothing is in progress, or planned, in the other sectors which would compromise the safety of personnel committed internally.)

To consider appointing a sector safety officer or officers, either for specific areas of concern (eg structure stability, dangerous terrain, etc.) or for general support. Such safety officers report direct to the Sector Commander, even if a "Safety Sector" has been designated, but must liaise with members of Safety Sector at every opportunity.

To confirm the Tactical Mode at 20 minute intervals.

A1.4.3 Crew Commanders

To monitor conditions in the risk area and draw the attention of the Sector Commander to significant developments.

To react immediately to adverse changes and withdraw crew members from the risk area without delay where necessary.

A1.4.4 Safety Sector (if operating)

- To survey operational sectors, identifying hazards, and advise the Sector Commander as appropriate.
- To liaise with sector safety officers, if appointed, to support and exchange information.
- To confirm the validity of the initial risk assessment and record as appropriate.
- To act as an extra set of eyes and ears to the Sector Commanders in monitoring the safety of personnel.

A1.5 Summary of the Procedure

Tactical Modes improve the safety of personnel on the incident ground.

There are only three Tactical Modes – **Offensive**, **Defensive**, or, **Transitional**.

Sectors can only be **Offensive** or **Defensive**. The incident can be **Offensive**, **Defensive** or, if changes or combinations of these two are in use, it will be **Transitional**.

The Incident Commander **must** formally adopt a Tactical Mode when operations are in progress.

When a Tactical Mode has been decided, the Incident Commander must ensure that everyone on the incident ground is aware of it.

Confirmation of the prevailing Tactical Mode must be maintained between Sector and Crew Commanders throughout the incident.

WEST YORKSHIRE FIRE SERVICE DYNAMIC RISK ASSESSMENT PROCESS

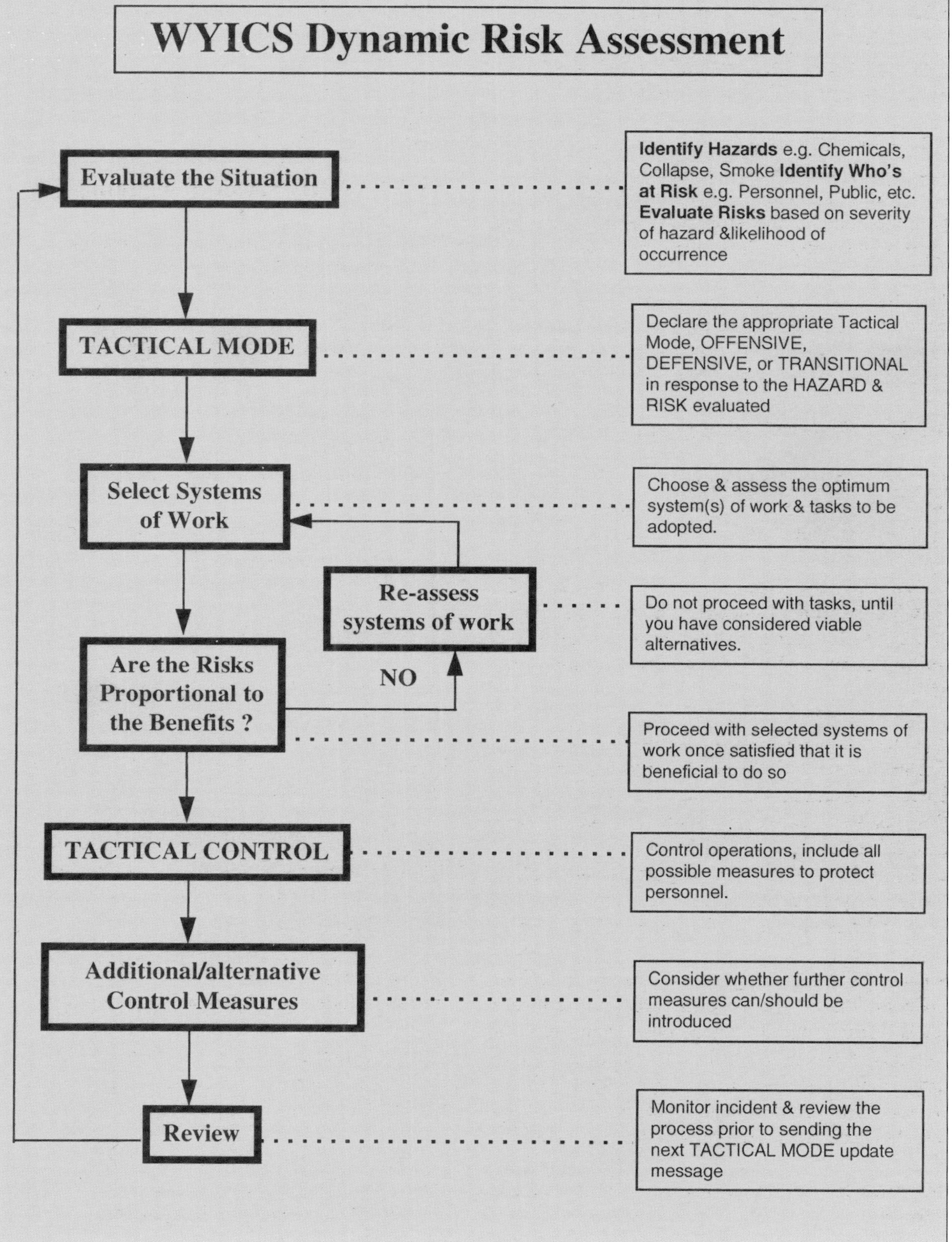

WYICS Dynamic Risk Assessment

Evaluate the Situation	**Identify Hazards** e.g. Chemicals, Collapse, Smoke **Identify Who's at Risk** e.g. Personnel, Public, etc. **Evaluate Risks** based on severity of hazard &likelihood of occurrence
TACTICAL MODE	Declare the appropriate Tactical Mode, OFFENSIVE, DEFENSIVE, or TRANSITIONAL in response to the HAZARD & RISK evaluated
Select Systems of Work	Choose & assess the optimum system(s) of work & tasks to be adopted.
Re-assess systems of work (NO → back to Select Systems of Work)	Do not proceed with tasks, until you have considered viable alternatives.
Are the Risks Proportional to the Benefits ?	Proceed with selected systems of work once satisfied that it is beneficial to do so
TACTICAL CONTROL	Control operations, include all possible measures to protect personnel.
Additional/alternative Control Measures	Consider whether further control measures can/should be introduced
Review (→ back to Evaluate the Situation)	Monitor incident & review the process prior to sending the next TACTICAL MODE update message

Figure A 1/1

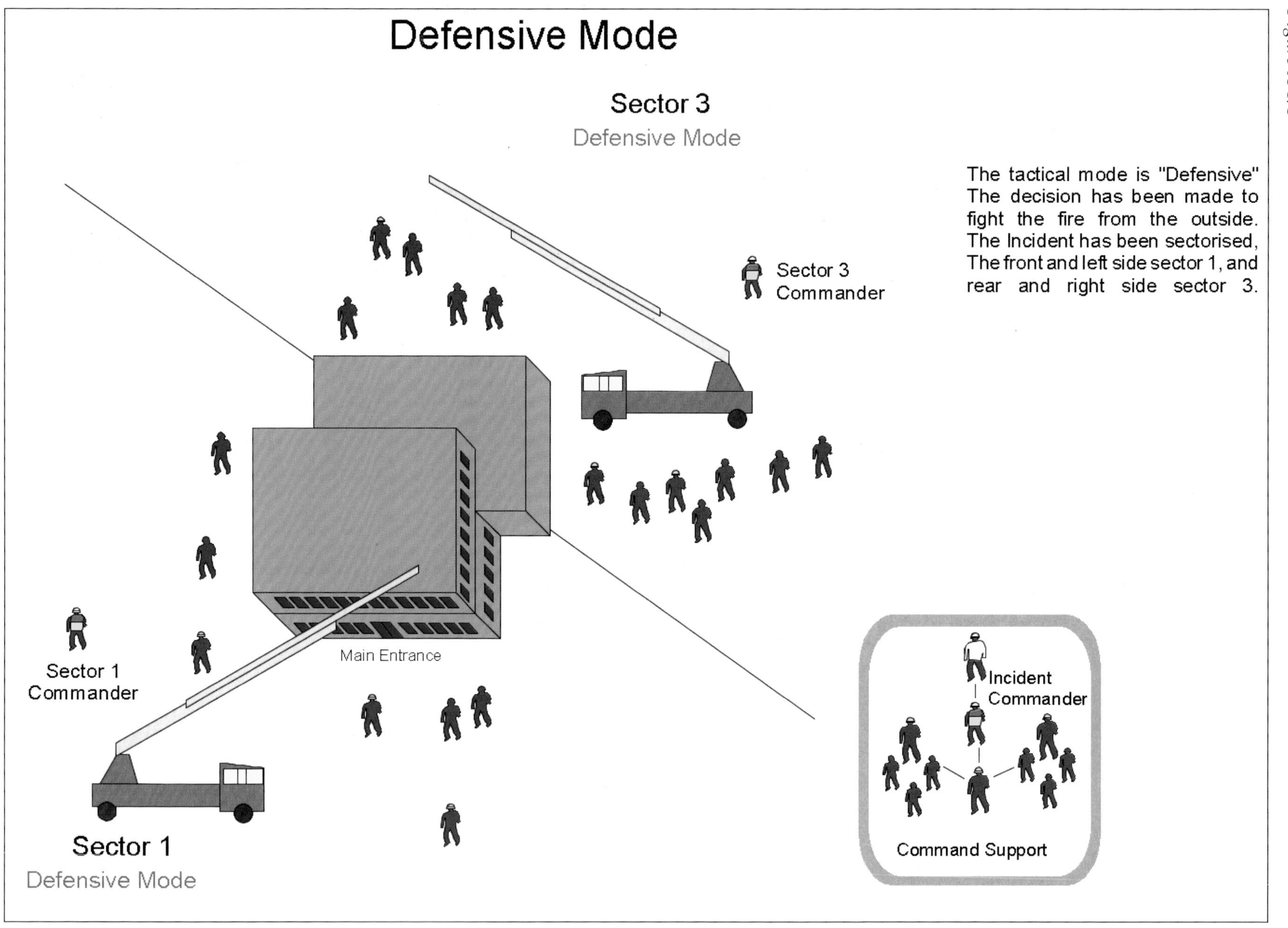

Defensive Mode
Sector 3
Defensive Mode
Sector 3 Commander
The tactical mode is "Defensive"
The decision has been made to fight the fire from the outside.
The Incident has been sectorised,
The front and left side sector 1, and rear and right side sector 3.
Main Entrance
Sector 1 Commander
Incident Commander
Command Support
Sector 1
Defensive Mode

Figure A 1/2

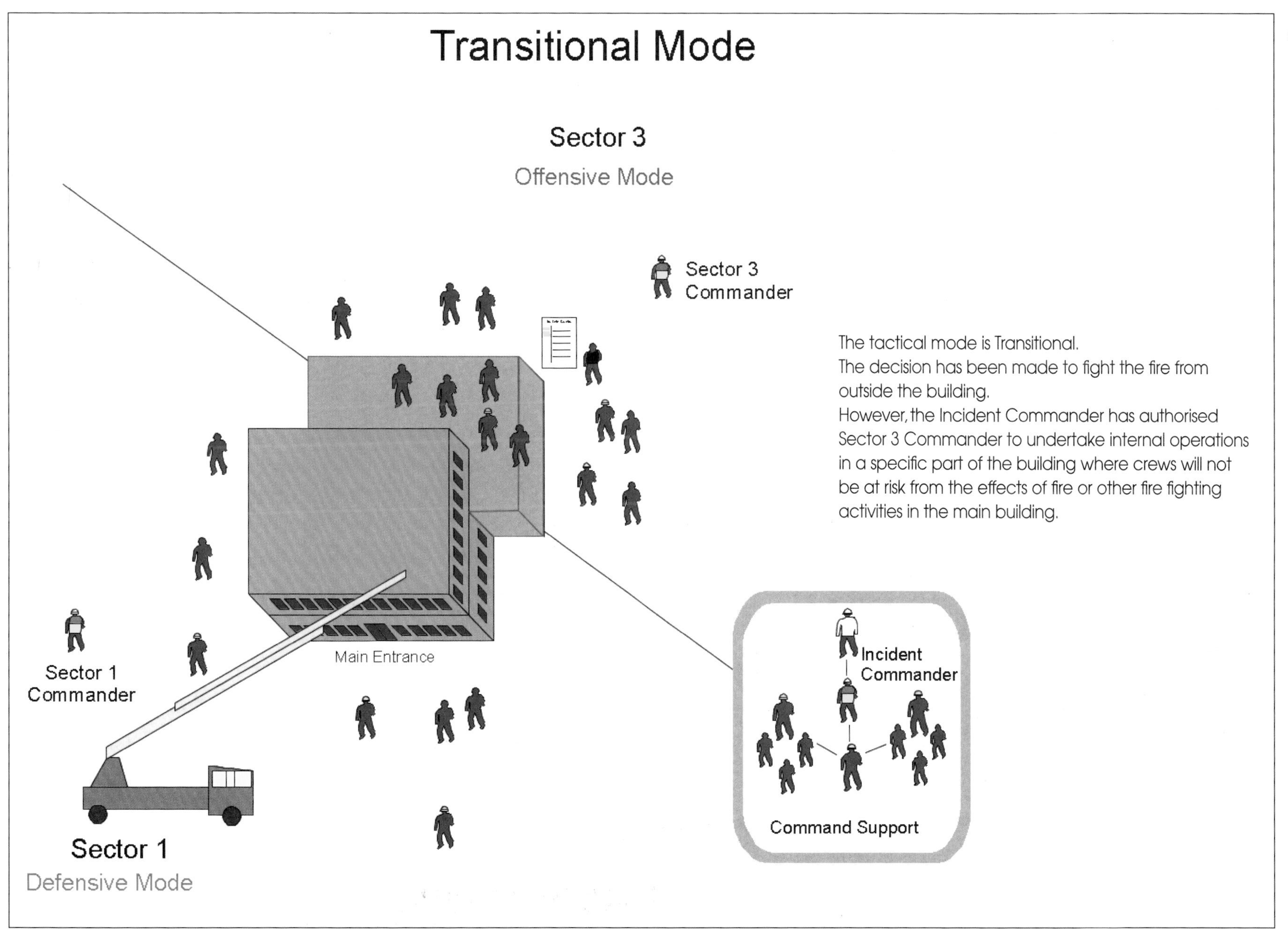
Transitional Mode
Sector 3
Offensive Mode
Sector 3 Commander
The tactical mode is Transitional.
The decision has been made to fight the fire from outside the building.
However, the Incident Commander has authorised Sector 3 Commander to undertake internal operations in a specific part of the building where crews will not be at risk from the effects of fire or other fire fighting activities in the main building.
Sector 1 Commander
Main Entrance
Incident Commander
Command Support
Sector 1
Defensive Mode

Figure A 1/3

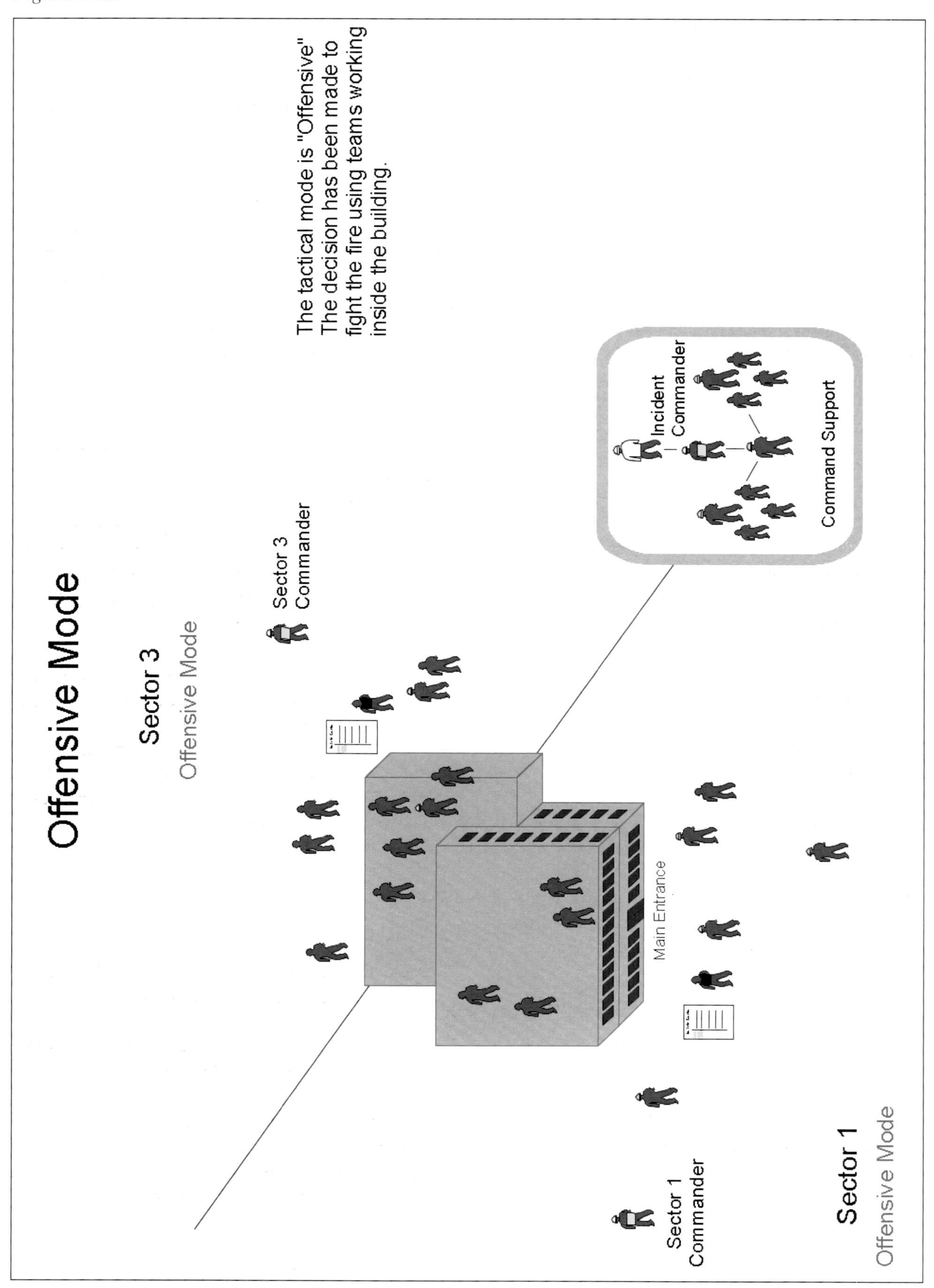

Offensive Mode
Sector 3
Offensive Mode
Sector 3
Commander
The tactical mode is "Offensive"
The decision has been made to
fight the fire using teams working
inside the building.
Incident
Commander
Command Support
Main Entrance
Sector 1
Commander
Sector 1
Offensive Mode

Figure 9 shows "West Yorkshire's Operational Risk Assessment Process"

Figures 10 to 12 give examples of the three Tactical modes.

Insert new diagram -----=on disc WYICS>DOC

Figure 9

A1.6 Further Guidance on Tactical Mode, with Examples

General Principles

Deciding not to commit a crew to an area of the incident because a risk assessment suggests the associated risk outweighs the potential benefit, is a defensive action.

So:

- Withdrawing a crew from a hazardous area because the risk has increased is a **defensive** action.
- Using jets from outside a building is a **defensive** action.
- Standing by awaiting expert advice, before committing crews is a **defensive** action.
- Standing by awaiting specialist equipment is a **defensive** action.

Deciding to commit a crew into a relatively hazardous area because the potential benefit outweighs the assessed risk, is an offensive action.

So:

- Committing BA crews to a smoke filled or toxic atmosphere is an **offensive** action.
- Committing crews to a structural collapse rescue is an **offensive** action.
- Committing crews to an RTA rescue is an **offensive** action.
- Committing a crew to fight a field fire is an **offensive** action.

If the incident is not sectorised, the assessments should be made on a crew by crew basis by the Incident Commander.

If such an assessment results in crews being committed to one area of an incident while being held back from other areas, then the incident is in **transition**. Before allowing operations which will cause the incident to become transitional the Incident Commander must be satisfied that the actions of one crew will not adversely affect others.

If the incident is sectorised, the tactical mode assessment must be made at sector level. The mode in a sector must be either offensive or defensive.

The mode in a sector will be proposed by the Sector Commander and confirmed by the Incident Commander. The mode may be subsequently changed only in consultation with the Incident Commander.

If, while in Offensive mode, there is a perceived need to withdraw crews from a hazardous area to ensure their safety this should be done immediately, i.e. defensive actions may be undertaken when in offensive mode, but should immediately be followed by consultation with the Incident Commander with a view to changing to Defensive Mode.

Examples

Example 1
3 pump house fire. Ground floor well alight, persons reported, believed to be in a first floor bedroom.
Large jet to work through a front window to knock down the fire on the ground floor.

2 BA teams committed from the rear door up stairs to search the first floor.
Incident is not sectorised
The incident is in Transitional Mode.

Later ...
Fire on the ground floor has been knocked down. BA team with hose reel enter ground floor to continue fire fighting.
The incident is in Offensive Mode.

Example 2
2 pump RTA persons trapped. Crews are working on the vehicles to effect rescues.
Incident is not sectorised.
The incident is in Offensive Mode.

Example 3
2 pump grass fire railway embankment involved. Crews standing by awaiting confirmation of caution passed to rail operator. No personnel have been committed to the embankment. No other operations are under way.
Incident is not sectorised.
The incident is in Defensive Mode.

Later...
Caution has been confirmed and lookouts are in place.
Crews are working on the embankment.
The incident is in Offensive Mode.

Example 4
2 pump RTA chemical tanker involved, the tanker is leaking a hazardous substance. No persons reported.
Road closed. Crews are standing by awaiting attendance of a specialist advisor and second tanker for decanting.
Incident is in Defensive Mode.

Later...
A crew has been committed in chemical protection suits to prevent the substance entering a drain. No operations at the crash scene
The incident is in transitional Mode.

Example 5
5 pump retail unit fire in a covered shopping mall. The retail unit is heavily involved in fire, all persons are accounted for. Smoke is issuing from the front of the unit into the shopping mall but is being contained and vented from a large atrium roof space. The smoke level is several metres above the mall floor and is stable. Operations in the mall are taking place in fresh air and within easy reach of final exits. The back of the unit is outside the mall. Smoke is issuing from the unit's roof and from an open loading bay.

Crews are at work inside the mall with jets into the front of the retail unit. Crews are at work at the rear of the unit with jets through the loading bay. No crews have made an entry to the retail unit.
The incident is in Defensive Mode.

(See figure A 1/4)

As a general guide in these circumstances, if conditions within a large building allow a sector or incident commander and associated staff to work within the building, then the risk assessments should be made on the basis of specific areas or compartments within the building rather than the whole building. Commanders and support staff should always work from an area of relative safety, so only crews committed beyond that area into a more hazardous environment could be considered as being committed offensively.

This is very similar to the principle of using a 'bridgehead' two floors below the fire floor of a multi-storey building for rigging and committing breathing apparatus wearers.

Figure A 1/4

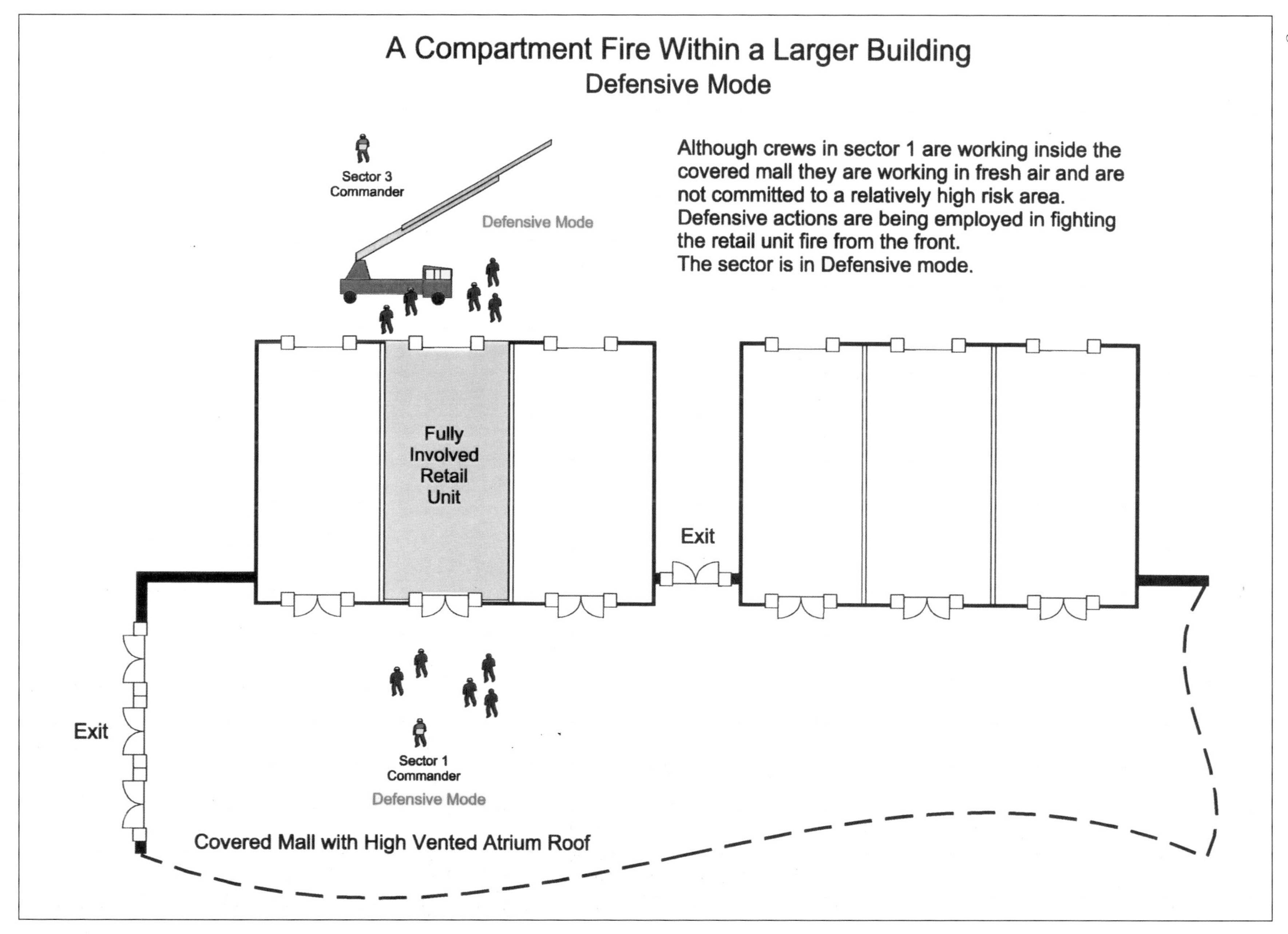
A Compartment Fire Within a Larger Building
Defensive Mode
Sector 3 Commander
Defensive Mode
Although crews in sector 1 are working inside the covered mall they are working in fresh air and are not committed to a relatively high risk area.
Defensive actions are being employed in fighting the retail unit fire from the front.
The sector is in Defensive mode.
Fully Involved Retail Unit
Exit
Exit
Sector 1 Commander
Defensive Mode
Covered Mall with High Vented Atrium Roof

APPENDIX 2 – Roles of the other Emergency Services

The roles detailed below reflect national guidance and are generally accepted by most services. There may be some local variations and where this occurs all services should be aware of them. E.g. in Scotland, police do not (all) use the Gold, Silver and Bronze system of command and control.

A2.1 Role of the Police at an Incident

The Police have their own policies and procedures for operational command.

The primary areas of Police responsibility, when attending an incident, can be summarised as follows:

- The saving of life in conjunction with other emergency services.
- Co-ordination of the emergency services and other subsidiary organisations.
- The protection and preservation of the scene.
- The investigation of the incident, in conjunction with other investigative bodies where applicable.
- The collation and dissemination of casualty information.
- Identification of victims on behalf of the Coroner who is the principal investigator when fatalities are involved.
- The restoration of normality at the earliest opportunity.

While the Police have the overall co-ordinating role, referred to as Police primacy, it is important to note that no single organisation has the sole responsibility for 'Command' at a large or major incident. The Fire Service will be expected to exercise some control over other emergency services at incidents involving fire, rescue and hazardous materials. At incidents other than fire, the police role is primarily one of co-ordination and facilitation of the overall situation. In such circumstances, the police will co-ordinate the response of the emergency services and facilitate the mobilisation and access of resources to the incident site. Where the Fire Service responds to a hazardous area to undertake its rescue role, the Police will normally set up and maintain an inner cordon while the Fire Service take charge of operations inside it.

A2.2 Police Incident Command Structure

Dependent on the size and location of the incident, three levels of police command may be implemented.

A2.3 Police Forward Control Point (Bronze or Operations Command)

Normally the first control to be established, or the nearest to the scene of the incident and responsible for immediate deployment and security. Initially under the command of the **Police Incident Officer**, the functions of the Forward Control Point may vary considerably dependent upon the type of incident, setting up arrangements and location of the Incident Control Post. Initially, the first police vehicle at the scene will serve as the Forward Control Point/Incident Control Post with the first officer on the scene acting as Incident Officer. His/her initial responsibility is to assume interim command, assess the situation and inform police control; but should not get involved in rescue work.

Where possible, all emergency service forward controls should be sited adjacent to one another but the fire service may influence their location in the interests of safety.

A2.4 Police Incident Control Post (Silver or Tactical Command)

If the size and/or nature of the incident requires it, a separate Police Incident Control Post will be set up to control, co-ordinate and manage the incident, providing a central point of contact for all emergency and specialist services. The Police Incident Control Post will be the responsibility of a co-ordinator and also under the command of the Police Incident Officer who will be senior in rank to the police officer initially having assumed command.

A2.5 Police Major Incident Control Room (Gold or Strategic Command)

The need for such a control is very much dependent on the size and scope of the incident. In some cases, even though there may be a number of casualties, all aspects of the operation can be co-ordinated through the Incident Control Post. However, with protracted incidents, where there are ongoing officer and logistical requirements, a Major Incident Control Room may be established to co-ordinate the overall response and provide facilities for Senior Command functions. This allows the Incident Control Post to fulfill its primary functions of co-ordinating and managing the operation within the 'controlled area' and requesting resources through the Major Incident Control Room. The Major Incident Control Room will usually be distant from the scene of operations, located either at the local Police Force Headquarters or a nominated police station. It will be under the command of the Overall Incident Commander, its routine would be the responsibility of a co-ordinator.

In summary:

- Police Gold is the overall incident commander located at the major incident control Room.
- Police Silver is the incident officer, located at the incident control post.
- Police Bronze is the sector commander(s) located at the forward control point(s).

A2.6 Role of the Ambulance Service at an Incident

The primary areas of Ambulance Service activity, when attending an incident, can be summarised as follows:

- Provide a focal point at the incident, through an Ambulance Control Point, for all Medical resources.
- The saving of life, in conjunction with other Emergency Services.
- The treatment and care of those injured at the scene, either directly or in conjunction with other medical personnel.
- Either directly, or in conjunction with medical personnel, determine the priority evacuation needs of those injured. (Triage)
- Determine the main Receiving and Supporting hospitals for the receipt of those injured.
- Arrange and ensure the most appropriate means of transporting those injured to the Receiving or Supporting hospitals.
- Ensuring that adequate medical staff and support equipment resources are available at the scene.
- The provision of communications facilities for National Health Service resources at the scene.
- The restoration to normality at the earliest possible opportunity.

A2.7 Ambulance Service Incident Command Structure

The Ambulance Service, like the Police, employ a three tier approach to Incident Command; these tiers are known as **Gold**, **Silver** and **Bronze**, although the role of Gold Command differs slightly from that of the Police.

A2.8 Ambulance Forward Control Point (Bronze Command)

Normally the first control to be established, or the nearest to the scene, where the Incident Officer/Forward Incident Officer can direct the operation with mobile communications. The Forward Control will also act as a focal point for the NHS/Medical resources at the initial point of patient contact on the scene. There may be a requirement for more than one Forward Control which will be sited outside the Inner Cordon. The access of Ambulance staff to the Inner Cordon will be controlled by the Fire Service.

A2.9 Ambulance Control Point (Silver Command)

An emergency control vehicle, readily identified by a green flashing light, providing an on-site communications facility which may be distant from the incident. It is to this location that all NHS/Medical resources should report and from where the Incident Officer will operate. Ideally this point should be in close proximity to the Police and Fire Service Control vehicles, subject to radio interference constraints.

A2.10 Ambulance Control Management (Gold Command)

Based in Headquarters Ambulance Control, the Ambulance Control Management Officer should not be involved directly with the controlling of the Ambulance Service resources but rather have a listening brief, providing an overview of how the incident is progressing. Through this monitoring, he/she will provide a valuable backup to the Ambulance Incident Officer, highlighting any likely problem areas and taking account of the implications for normal day to day operations. He/she should be responsive to the needs of the Ambulance Incident Officer at the scene.

In summary:

- Ambulance Gold is the ambulance control management officer located at ambulance headquarters control.
- Ambulance Silver is the incident officer located at the ambulance control point.
- Ambulance Bronze is the forward incident officer located at the forward control point.

APPENDIX 3 – The Psychology of Command

The psychology of command is beginning to emerge as a distinct research topic for psychologists interested in selection, training, competence assessment, decision making, stress management, leadership and team working. The following overview of recent research into decision making, stress and leadership is based on Flin (1996) which gives a more detailed examination of these issues.

A3.1 Decision Making

The decision making skill of the Incident Commander is one of the essential components of effective command and control in emergency response. Despite the importance of high speed decision making in the fire service and a number of other occupations, it has only been very recently that research psychologists have begun to investigate leaders' decision making in demanding, time-pressured situations.

The traditional decision-making literature from management, statistics and economics is very extensive but it offers little of relevance to the Incident Commander, as it tends to be derived from studies of specified problems (often artificial in nature), inexperienced decision makers and low stake payoffs. Moreover, it is rarely concerned with ambiguous, dynamic situations, life threatening odds, or high time pressure, all important features of a fire or rescue environment.

If we turn to the standard psychological literature on decision-making it tells us almost nothing of emergency decision making, as so much of it is based on undergraduates performing trivial tasks in laboratories. Similarly, the management research is concerned with individuals making strategic decisions when they have several hours or days to think about the options, carefully evaluating each one in turn against their business objectives using decision analysis methods. These provide a range of explanatory frameworks which may have value for managers' decision making where they are encouraged to emulate an analytical style of decision making. At its simplest form this usually incorporates the following stages.

(1) Identify the problem.

(2) Generate a set of options for solving the problem/choice alternatives.

(3) Evaluate these options concurrently using one of a number of strategies, such as weighting and comparing the relevant features of the options.

(4) Choose and implement the preferred option.

In theory, this type of approach should allow you to make the 'best' decision, provided that you have the mental energy, unlimited time and all the relevant information to carry out the decision analysis. This is typically the method of decision making in which managers are trained. But we know from our everyday experience that when we are in a familiar situation, we take many decisions almost automatically on the basis of our experience. We do not consciously generate and evaluate options, we simply know the right thing to do. This may be called intuition or 'gut feel' but, in fact, to achieve these judgements some very sophisticated mental activity is taking place. So we can compare these two basic types of decision making, the slower but more analytic comparison and the faster, intuitive judgement. Which style do commanders use when deciding what to do at the scene of an incident?

A3.2 Naturalistic Decision Making (NDM)

In the last ten years there has been a growing interest by applied psychologists into naturalistic decision making (NDM) which takes place in complex real world settings (Klein et al, 1993; Zsambok & Klein, 1997; Flin et al, 1997). These researchers typically study experts' decision making in dynamic environments such as flight decks, military operations, firegrounds, hospital trauma centres/intensive care units and high hazard industries, for example nuclear plant control rooms. This NDM research has enormous significance for the understanding of how commanders and their teams make decisions at the scene of an incident as it offers descriptions of what expert commanders actually do when taking operational decisions in emergencies.

Ten factors characterize decision making in naturalistic settings:

(1) Ill defined goals and ill structured tasks.

(2) Uncertainty, ambiguity and missing data.

(3) Shifting and competing goals.

(4) Dynamic and continually changing conditions.

(5) Action feedback loops (real-time reactions to changed conditions).

(6) Time stress.

(7) High Stakes.

(8) Multiple players (team factors).

(9) Organizational goals and norms.

(10) Experienced decision makers

In typical NDM environments information comes from many sources, is often incomplete, can be ambiguous, and is prone to rapid change. In an emergency, the Incident Commander and her or his team are working in a high stress, high risk, time pressured setting and the lives of those affected by the emergency, (including their own fire rescue personnel) may be dependent on their decisions.

How then do they decide the correct courses of action? In the view of the NDM researchers, traditional, normative models of decision making which focus on the process of option generation and simultaneous evaluation to choose a course of action do not frequently apply in NDM settings. There are a number of slightly different theoretical approaches within the NDM fraternity to studying decision making but they all share an interest in dynamic high pressure domains where experts are aiming for satisfactory rather than optimal decisions due to time and risk constraints.

A3.3 Recognition-Primed Decision Making (RPD)

Dr Gary Klein of Klein Associates, Ohio, conducts research into decision making by attempting to 'get inside the head' of decision makers operating in many different domains. Klein's approach stemmed from his dissatisfaction with the applicability of traditional models of decision making to real life situations, particularly when the decisions could be lifesaving. He was interested in operational environments where experienced decision makers had to determine a course of action under conditions of high stakes, time pressures, dynamic settings, uncertainty, ambiguous information and multiple players.

Klein's research began with a study of urban fireground commanders who had to make decisions such as whether to initiate search and rescue, whether to begin an offensive attack or concentrate on defensive precautions and how to deploy their resources (Klein et al, 1986) They found that the fireground commanders' accounts of their decision making did not fit in to any conventional decision-tree framework.

"The fireground commanders argued that they were not 'making choices', 'considering alternatives', or 'assessing probabilities'. They saw themselves as acting and reacting on the basis of prior experience; they were generating, monitoring, and modifying plans to meet the needs of the situations. Rarely did the fireground commanders contrast even two options. We could see no way in which the concept of optimal choice might be applied. Moreover, it appeared that a search for an optimal choice could stall the fireground commanders long enough to lose control of the operation altogether. The fireground commanders were more interested in finding actions that were workable, timely, and cost-effective." (Klein et al, 1993, p139).

During post-incident interviews, they found that the commanders could describe other possible courses of action but they maintained that during the incident they had not spent any time deliberating about the advantages or disadvantages of these different options.

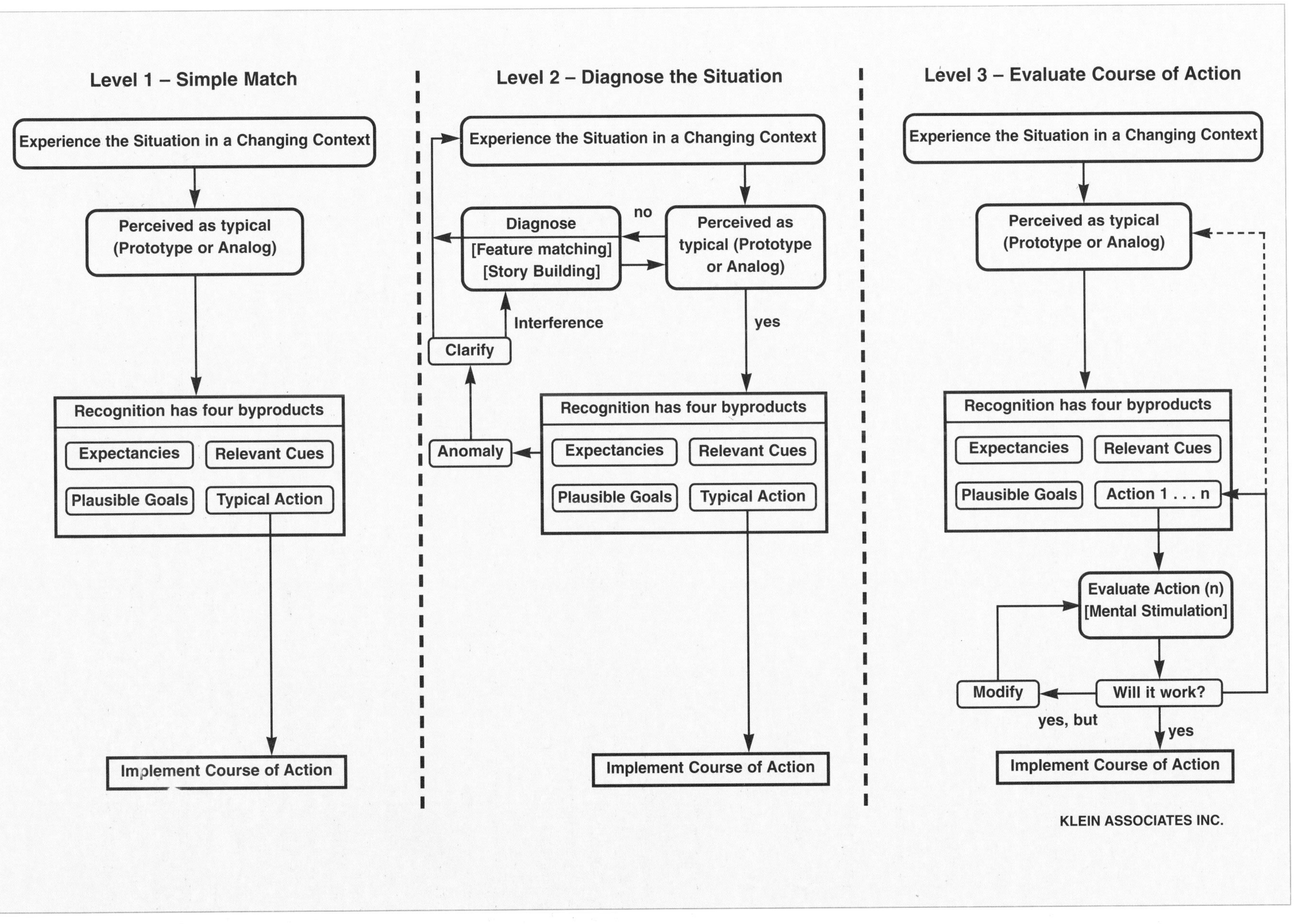

Figure A3/1 The Recognition Primed Decision Model (Klein, 1996) Reproduced with permission of Klein Associates.

It appeared that these Incident Commanders had concentrated on assessing and classifying the situation in front of them. Once they recognised that they were dealing with a particular type of event, they usually also knew the typical response to tackle it. They would then quickly evaluate the feasibility of that course of action, imagining how they would implement it, to check whether anything important might go wrong. If they envisaged any problems, then the plan might be modified but only if they rejected it, would they consider another strategy.

Klein Associates have also studied other decision makers faced with similar demand characteristics (e.g., tank platoon captains, naval warfare commanders, intensive care nurses) and found the same pattern of results. On the basis of these findings they developed a template of this strategy called the Recognition-Primed Decision Model. This describes how experienced decision makers can rapidly decide on the appropriate course of action in a high pressure situation.

The model has evolved into three basic formats (see Figure A 3/1).

In the simplest version, shown as **Level 1**, the decision maker recognizes the type of situation, knows the appropriate response and implements it.

If the situation is more complex and/or the decision maker cannot so easily classify the type of problem faced, then as in **Level 2**, there may be a more pronounced diagnosis (situation assessment) phase. This can involve a simple feature match where the decision maker thinks of several interpretations of the situation and uses key features to determine which interpretation provides the best match with the available cues. Alternatively, the decision maker may have to combine these features to construct a plausible explanation for the situation, this is called story building, an idea that was derived from legal research into juror decision making. Where the appropriate response is unambiguously associated with the situation assessment it is implemented as indicated in the Level 1 model.

In cases where the decision maker is less sure of the option, then the RPD model, **Level 3** version indicates that before an action is implemented there is a brief mental evaluation to check whether there are likely to be any problems. This is called mental simulation or preplaying the course of action (an 'action replay' in reverse) and if it is deemed problematical then an attempt will be made to modify or adapt it before it is rejected. At that point the commander would reexamine the situation to generate a second course of action.

Key features of the RPD model are as follows:

- Focus on situation assessment.
- Aim is to satisfy not optimize.
- For experienced decision makers, first option is usually workable.
- Serial generation and evaluation of options (action plans).
- Check action plan will work using mental simulation.
- Focus on elaborating and improving action plan.
- Decision maker is primed to act.

To the decision maker, the NDM type strategies (such as RPD) feel like an intuitive response rather than an analytic comparison or rational choice of alternative options. As 'intuition' is defined as, *"the power of the mind by which it immediately perceives the truth of things without reasoning or analysis"* then this may be an acceptable label for RPD which is rapid situation assessment to achieve pattern recognition and associated recall of a matched action plan from memory.

At present this appears to be one of the best models available to apply to the emergency situation whether the environment is civilian or military; onshore or offshore; aviation, industrial, or medical. In the USA, the RPD model is being widely adopted, it is being used at the National Fire Academy as well as in a number of military, medical, aviation and industrial settings (see Klein, 1998). The RPD model and associated research techniques have begun to generate a degree of interest in the UK, most notably by the Defence Research Agency and the Fire Service.

A3.4 Command roles and decision style

Obviously the RPD approach is not appropriate for all types of operational decisions and other NDM researchers have been developing taxonomies of the different types of decisions other emergency commanders, such as pilots, make in different situations (see Figure A3/2). The NASA Crew Factors researchers (Orasanu, 1995) have found that two key factors of the initial situation assessment are judgements of time and risk and that these may determine the appropriate decision strategy to use. The issue of dynamic risk analysis is a significant component of situation assessment on the fireground as discussed in Chapter 3 (see also *Fire Engineers Journal*, May, 1998).

If we consider the Orasanu model, the key skill is matching the correct decision style to the demands or allowances of the situation. For example not using the fast intuitive RPD style when there is time to evaluate options. Furthermore senior fire officers in strategic command roles may require special training to discourage them from using the fast RPD approach when a slower, analytical method would be more appropriate (Fredholm, 1997).

There are significant differences in the balance of cognitive skills required of commanders, depending on their role (rather than rank) in a given operation, ascending from operational or task level, to tactical command, to strategic command (Home Office, 1997). From studies of commanders' decision strategies (see Flin, 1996; Flin et al, 1997; Zsambok & Klein, 1997) these roles are briefly outlined below in terms of the decision skills required.

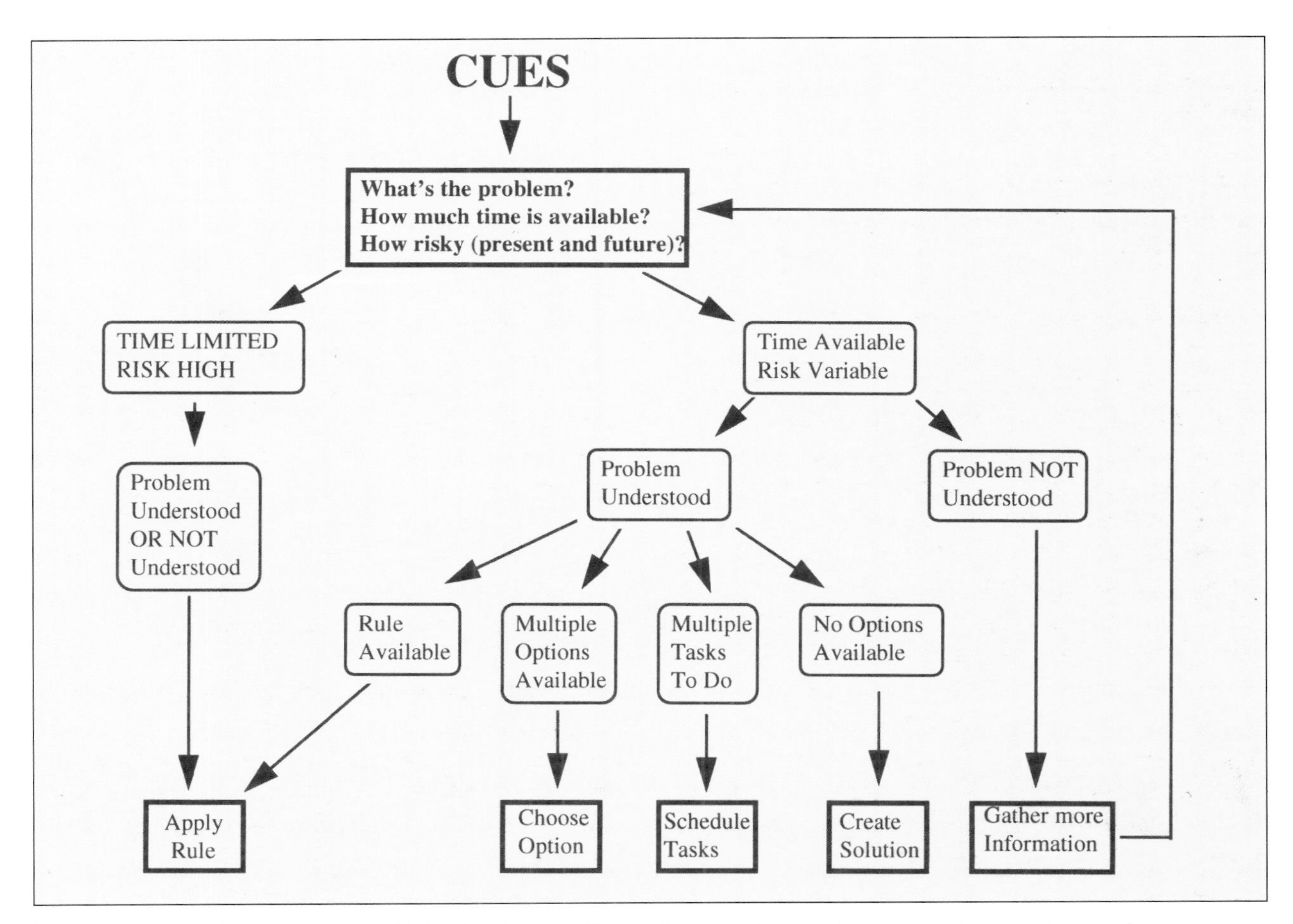

Figure A3/2 Decision process model for fixed wing pilots. (Orasanu 1995B) Reprinted with permission of the Human Factors and Ergonomics Society.

Strategic Command

This involves the overall policy of command and control, deciding the longer term priorities for tactical commanders and planning for contingencies depending on the enemy's response. The task also contains a strong analytical element, as co-ordination of multiple sources of information and resources demands an awareness that cannot be based on procedures alone.

The decision making style assumed to be adopted for strategic decision making is creative or analytical, since the situations encountered will feature a number of novel elements or developments the strategic commander has not previously encountered.

Neither time pressure nor high immediate risk should be influencing command at this level, where the aim, if possible, is to devise an optimal solution for the situation, taking into account the wider and longer term implications. The strategic commander is usually remote from the incident and will be supported throughout by a team of lower ranking officers.

Tactical Command

This refers to the planning and co-ordination of the actions determined at the strategic level. Due to the higher time pressure at this level, decision making is based to a much greater extent on condition-action matching, or rule-based reasoning. This style is characterised by controlled actions derived from procedures stored in memory. Control of behaviour at this level is goal oriented and structured by 'feed-forward control' through a stored rule. Stored rules are of the type if (state) then (diagnosis) or if (state) then (remedial action).

The tactical decision maker is likely to be on-scene, with a remit to maintain a good mental model of the evolving plan and unfolding events. Situation assessment is expected to be a more significant component of tactical decision making than spending time choosing appropriate responses. However the tactical commander may have to 'create' time to engage in reflective thinking and when necessary to use more analytic decision strategies to evaluate alternative courses of action.

Kerstholt (1997, p189) found from an interview study with battalion commanders of peace-keeping operations, that, *"decisions were mostly made analytically in the planning phase and intuitively during the execution of the mission. By analytic procedure we meant that several options were explicitly weighed against each other, whereas an intuitive decision meant that the commander immediately 'knew' which decision to take."*

Operational Command

This involves front line or sector commanders who have to implement orders from the tactical level. They are operating in real time and have to react rapidly to situational demands. Decision making at this level is assumed to contain rule-based and intuitive elements. It is assumed that under time pressure and at high risk, they primarily make decisions based on pattern recognition (e.g. RPD) of the situations encountered. Ongoing situation awareness must remain very high as their performance depends on rapid identification of the situation and fast access to stored patterns of pre-programmed responses.

Only when time permits will they be able to engage in analytic decision making and option comparison. Striving to find optimal solutions runs the risk of 'stalling' their decision making, therefore their main objective is to find a satisfactory, workable course of action.

A3.5 Styles of Command Decision Making

From the above description of decision making techniques associated with particular command roles, there appear to be four main styles of decision making used by commanders: **creative**, **analytical**, **procedural** and **intuitive**.

The most sophisticated (and resource intensive) is **creative** problem solving which requires a diagnosis of an unfamiliar situation and the creation of a novel course of action. This is the most demanding of the four techniques, requires significant expertise and as Kerstholt (1997) found, is more likely to be used in a planning phase rather than during an actual operation.

Decision Style	*Cognitive Processes*
Creative Problem Solving	Diagnosis of unfamiliar situation requiring extensive information search and analysis. Development/synthesis of new courses of action. Knowledge-based reasoning.
Analytical Option Comparison	Retrieval and comparison of several courses of action. High working memory load. Knowledge-based reasoning.
Procedural/ Standard Operating Procedures	Situation identification. Retrieval (and rehearsal) of rules for course of action. Explicit rule-based reasoning. If *x* then *y*.
Intuitive/ Recognition-primed decisions (RPD)	Rapid situation recognition based on pattern matching from long-term memory. Implicit rule based or skill-based. 'Gut feel'.

Command decision styles

Analytical decision making also requires a full situation assessment, rigorous information search and then recall, critical comparison and assessment of alternative courses of action. Again with proper preparation, some of these option choices may already have been evaluated during exercises or planning meetings. These are the two most powerful decision techniques as they operate on large information sets, but consequently they require far greater cognitive processing. Thus, they take a longer time to accomplish, and for most individuals can only be used in situations of relative calm and minimal distraction.

In fast moving, high risk situations these styles are difficult if not impossible to use, and in order to maintain command and control, officers have to rely on procedural or intuitive styles which will produce a satisfactory, if not an optimal decision.

Procedural methods involve the identification of the problem faced and the retrieval from memory of the rule or taught method for dealing with this particular situation. Such decision methods (e.g. drills, routines and standard procedures) are frequently practised in training.

With experience, officers may also use the fastest style of decision making, **intuitive** or recognition-primed decision making described above. In this case there may not be a written rule or procedure but the commander rapidly recognises the type of situation and immediately recalls an appropriate course of action, on the basis of prior experience.

The evidence suggests that commanders use all four decision styles to a greater or lesser degree depending on the event characteristics and resulting task demands. For more senior commanders, distanced from the front line, the task characteristics change in terms of time frame, scale, scope and complexity, necessitating greater use of analytical and creative skills (Fredholm, 1997).

Studies of military and aviation commanders have shown that the following factors are of particular significance in determining decision style:

- available time.
- level of risk.
- situation complexity/familiarity, (or none at all) .
- availability of information.

The training implications of applying this new decision research to fire and rescue operations is first to determine the types of situations where experienced fire commanders use the intuitive RPD type of decision making. In these situations the critical focus will be on situation assessment. So the next stage is to discover the cues these experts use when quickly sizing up an incident and the responses they would choose to apply once they have assessed the situation.

Less experienced commanders need to be trained to recognize these key features or cues of different scenarios using simulated incidents with detailed feedback on their decision making. They need to develop a store of incident memories (from real events, simulator training, case studies, expert accounts) which they can use to drive their search for the critical classifying information at a new incident.

The US Marines who favour the RPD model have developed a very useful volume of 15 decision exercises in *Mastering Tactics: A Tactical Decision Games Workbook* (Schmitt, 1994, see Klein, 1998). These are a series of tactical decision scenarios where a description of a problem is presented and officers are required to quickly work out and explain a solution to the problem which can then be discussed with the team and/or the trainer. This assists officers to learn the critical cues for given types of situations and to store methods of dealing with new situations.

In essence the basis of good command training must be a proper understanding of the decision making processes utilised by effective commanders.

Psychologists can offer a range of research techniques to begin to explore in a more scientific fashion the skills of incident command (e.g. Burke, 1997; Flin et al, 1997). For instance, one of the most salient features of a fireground commander's decision task is the speed of fire development. Brehmer (1993) is particularly interested in this type of dynamic decision task, which he believes has four important characteristics: a series of decisions, which are interdependent, a problem which changes autonomously and as a result of the decision maker's actions, and a real time scenario.

He gives the following example, *"Consider the decision problems facing a fire chief faced with the task of extinguishing forest fires. He receives information about fires from a spotter plane, and on the basis of this information, he then sends out commands to his firefighting units. These units then report back to him about their activities and locations as well as about the fire, and the fire chief uses this information (and whatever other information he may be able to get, e.g., from a personal visit to the fire and the fire fighting units) to issue new commands until the fire has been extinguished."* (p1).

Brehmer and his colleagues have developed a computer programme (FIRE) based on a forest fire scenario which incorporates the four elements of dynamic decision making described above. The decision maker takes the role of the fire chief and using the grid map of the area shown on the computer screen, she or he has to make a series of decisions about the deployment of fire fighting resources with the goal of extinguishing the fire and protecting a control base.

The commander's actions are subject to feedback delays, that is time delay in actions being implemented or in the commander receiving status update information. Brehmer's studies have shown that decision makers frequently do not take such feedback delays into account, for example sending out too few firefighting units because they do not anticipate that the fire will have spread by the time they receive the status report.

He argues that the decision maker needs to have a good 'mental model' of the task in order to control a dynamic event, such as a forest fire, and his research has enabled him to identify several problems of model formation: dealing with complexity, balancing competing goals, feedback delays and taking into account possible side effects of actions. Brehmer (1993) uses control theory to encapsulate the dynamic decision process, *"the decision maker must have clear goals, he must be able to ascertain the state of the system that he seeks to control, he must be able to change the system, and he must have a model of the system."* (p10).

A3.6 Causes of Stress for Commanders

In fireground operations, stress may also have an impact on commanders' decision making and techniques for managing this need to be considered (see Flin 1996 for further details).

The effects of stress on commanders' thinking and decision making ability are of particular interest. Charlton (1992) who was responsible for the selection of future submarine commanders referred to the '**flight**, **fight** or **freeze**' response manifested as problems in decision making, '**tunnel vision**', **misdirected aggression**, **withdrawal**, and the '**butterfly syndrome**' *"where the individual flits from one aspect of the problem to another, without method solution or priority"* (p54). He also mentions self delusion where the student commander denies the existence or magnitude of a problem, regression to more basic skills, and inability to prioritise.

Weiseath (1987) discussing the enhanced cognitive demands for leaders under stress describes reduced concentration, narrowing of perception, fixation, inability to perceive simultaneous problems, distraction, difficulty in prioritising and distorted time perception.

Brehmer (1993) argues that three 'pathologies of decision making' can occur, he calls these

(i) **thematic vagabonding** – when the decision maker shifts from goal to goal;

(ii) **encystment** – the decision maker focuses on only one goal that appears feasible, and as in (i) fails to consider all relevant goals; and

(iii) a **refusal** to make any decisions.

Not all researchers agree that the decision making of experienced Incident Commanders will be degraded by exposure to acute stressors. Klein (1998) points out that these effects are most typical when analytical decision strategies are used, in contrast, the recognition-primed type of decision strategy employed by experts under pressure may actually be reasonably stress-proof.

A3.7 Leadership

Leadership ability is generally deemed to be a key attribute of an Incident Commander and to some extent may be regarded as an umbrella term for the required competencies which have to be trained. However, finding a precise specification of the required behaviours or the style of leadership is rather less frequently articulated.

Leadership within a military context embodies the concepts of command, control, organization and duty. There has been extensive military research into leadership much of which unfortunately, never sees the light of day outside the defence research community.

The dominant model of leadership for training in the British armed services, the emergency services and in lower level management, is Adair's (1988) Action Centred Leadership with its simple three circles model.

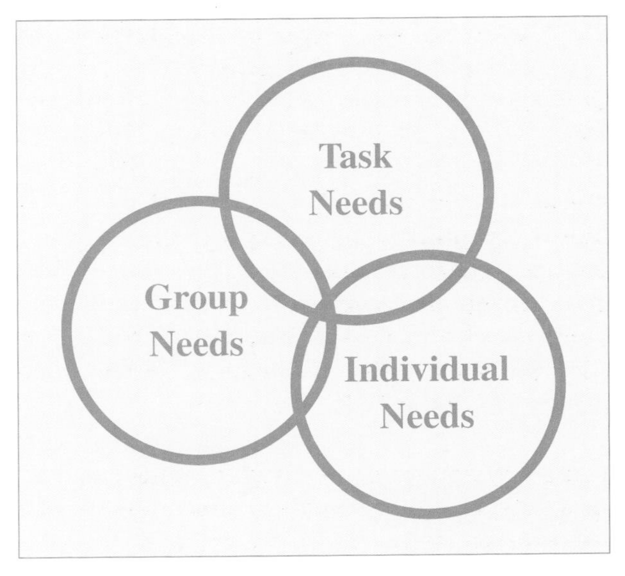

Figure A.3/3 showing Adair's Leadership Model

Adair developed his ideas from his experiences with the British Army, and he maintained that the effective leader must focus on the needs of the individual, the task and the team. This functional model has not changed significantly since its initial exposition thirty years ago and continues to be taught in a wide range of management courses. While the three circles diagram and the associated advice to leaders is intuitively appealing, there has been little empirical work to test whether it can actually function as an explanatory theory of leadership in routine managerial duties or emergency command situations.

The managerial research literature on leadership is a progression from a long standing focus on leadership characteristics, to research in the 1960s on leader behaviours (e.g. autocratic *vs* democratic; team *vs* task), to an awareness that "one size fits all" recommendations of the best leadership style are unlikely to work. The contingency theories emphasised that leadership style cannot be considered in isolation. Thus, what is effective leadership behaviour is likely to be dependent on the leader's personality and skills, the situation, and the competence and motivation of the group being led. Thus the most effective leader needs to:

(i) be able to diagnose the situation (the task/problem, the mood, competence, motivation of the team),

(ii) have a range of styles available (e.g. delegative, consultative, coaching, facilitating, directive),

(iii) match her or his style to the situation (for example Hersey and Blanchard's (1988) model of situational leadership).

In an emergency which has high time pressure and risk, then it is unlikely that a consultative leadership style would be totally appropriate and while the Incident Commander needs to solicit advice from available experts and to listen to the sector commanders, the appropriate style is likely to be closer to directive than democratic.

The need for a perceptible change in leadership style is very obvious when observing simulated emergency exercises when the time pressure and task demands are increased. Moreover, this sends a very important message to the rest of the team that the situation is serious and that they will also have to 'change gear' and sharpen their performance.

Within the business world, the current fashions in leadership style are the delegative, consultative styles, couched in notions of empowerment and transformational leadership. These approaches have not been developed with the Incident Commander in mind, and while it was argued above that a consultative style may be inappropriate, particularly in the opening stages of an incident, this does not mean that there should be no delegation to more junior commanders.

In a larger incident considerable authority has to be devolved to sector commanders who will be required to take critical decisions and who will not always have time or opportunity to seek the opinion of the Incident Commander. These individuals need to have the expertise and the confidence to make decisions as the need arises.

The essential point is that the commander should be comfortable with the style required and that the front-line commanders should have a clear understanding of their delegated authority and the Incident Commander's plan of action.

Finally, the Incident Commander does not, and should not work alone. The need for effective team performance on the incident ground remains paramount. Recent advances in team training, known as Crew Resource Management (CRM) have been developed by the aviation industry and are now used in medicine and the energy industry. The focus is on non-technical skills relevant to incident command, such as leadership, situation awareness, decision making, team climate and communication (see Flin, 1995b; Salas et al, in press for further details). Fire officers who have studied this particular type of human factors training have argued that it has clear applications for the fire service (Bonney, 1995, Wynne, 1994).

References

Brunacini, A. (1991) Command safety: A wake-up call. *National Fire Protection Association Journal*, January, 74–76.

Burke, E. (1997) Competence in command: Research and development in the London Fire Brigade. In R. Flin, E. Salas, M. Strub & L. Martin (Eds) *Decision Making under Stress.* Aldershot: Ashgate.

Driskell, J. & Salas, E. (1996) (Eds) *Stress and Human Performance*. Mahwah, NJ: LEA.

Flin, R. (1995a) Incident command: Decision making and team work. *Journal of the Fire Service College, 1*, 7–15.

Flin, R. (1995b) Crew Resource Management for teams in the offshore oil industry. *Journal of European Industrial Training, 19,9*, 23–27.

Flin, R. (1996) *Sitting in the Hot Seat: Leaders and Teams for Critical Incident Management.* Chichester: Wiley.

Flin, R., Salas, E., Strub, M. & Martin, L. (1997) (Eds) *Decision Making under Stress: Emerging Themes and Applications*. Aldershot: Ashgate.

Fredholm, L. (1997) Decision making patterns in major fire-fighting and rescue operations. In R. Flin, E. Salas, M. Strub & L. Martin (Eds) *Decision Making under Stress*. Aldershot: Ashgate.

Home Office (1997) *Dealing with Disaster.* Third edition. London: TSO

Klein, G. (1998) *Sources of Power. How People Make Decisions*. Cambridge, Mass: MIT Press.

Klein, G. (1997) The Recognition-Primed Decision (RPD) model: Looking back, looking forward. In C. Zsambok & G. Klein (Eds) *Naturalistic Decision Making*. Mahwah, NJ: Lawrence Erlbaum.

Klein, G., Calderwood, R., & Clinton-Cirocco, A. (1986) Rapid decision making on the fireground. *In Proceedings of the Human Factors Society 30th Annual Meeting*. San Diego: HFS.

Klein, G., Orasanu, J., Calderwood, R. & Zsambok, C. (1993). (Eds.) *Decision Making in Action*. New York: Ablex.

Murray, B. (1994) More guidance needed for senior commanders on the fireground. *Fire, 87*, June, 21-22.

Orasanu, J. & Fischer, U. (1997) Finding decisions in naturalistic environments: The view from the cockpit. In C. Zsambok & G. Klein (Eds) *Naturalistic Decision Making*. Mahwah, NJ: LEA.

Salas, E., Bowers, C. & Edens, E. (in press) (eds.) *Applying Resource Management in Organisations*. New Jersey. LEA.

Schmitt, J. (1994) *Mastering Tactics. Tactical Decision Game Workbook*. Quantico, Virginia. US Marine Corps Association.

Zsambok, C. & Klein, G. (1997) (Eds) *Naturalistic Decision Making*. Mahwah, NJ: LEA.

APPENDIX 4 – Legal Considerations in Command & Control: Legislative Requirements

The responsibility placed on Incident Commanders by Section 1 The Fire Services Act 1947 (as amended), has increased by subsequent legislation or civil law cases since the 1947 Act.

Of major impact has been the introduction of the Health and Safety at Work etc. Act 1974, hereto referred as H.A.S.A.W.A, and subordinate legislation such as the Management of Health and Safety at Work Regulations 1992. To support legal requirements for health and safety of employees and other persons at incidents.

In order to achieve this during any incident, the Incident Commander requires a knowledge of health and safety responsibilities and fireground systems to support, so far as is reasonable, legal requirements.

A4.1 Fire Services Act 1947 (as amended)

Section 1

(1) It shall be the duty of every Fire Authority in Great Britain to make provision for firefighting purposes, and in particular every Fire Authority shall secure:

(a) The services for their area of such a fire brigade and such as may be necessary to meet efficiently all normal requirements.

(b) The efficient training of the members of the fire brigade.

(c) Efficient arrangements for dealing with calls for the assistance of the fire brigade in case of fire and for summoning members of the brigade.

(d) Efficient arrangements for obtaining, by inspection or otherwise, information required for firefighting purposes with respect to the character of the buildings and other property in the area of the Fire Authority, the available water supplies and the means of access thereto and other material local circumstances.

(e) Efficient arrangements for ensuring that reasonable steps are taken to prevent or mitigate damage to property resulting from measures taken in dealing with fires in the area of the Fire Authority.

(f) Efficient arrangements for the giving, when requested, of advice in respect of buildings and other property in the area of the Fire Authority as to fire prevention, restricting the spread of fires, and means of escape in case of fire.

Section 2 – Arrangements for Mutual Assistance

(1) It shall be the duty of Fire Authorities, so far as practicable, to join in the making of schemes (hereafter in this section referred to as 'reinforcement schemes') for securing the rendering of mutual assistance for the purpose of dealing with fires occurring in the areas of Authorities participating in a reinforcement scheme where either:

(a) It is necessary to supplement the services provided under the last foregoing section by the Authority in whose area the fire occurs, or

(b) Reinforcements at any fire can be more readily obtained from the resources of other authorities participating in the scheme than from those of the authority in whose area the fire occurs.

(8) A Fire Authority may enter into arrangements with persons (not being other Fire Authorities) who maintain fire brigades to secure, on such terms as to payment or otherwise as may be provided by or under the arrangements, the provision by those persons of assistance for the purpose of dealing with fires occurring in the area of the Authority where either:

(a) It is necessary to supplement the services provided by the Authority under the last foregoing section, or

(b) Reinforcements at any fire occurring in the area of the Authority can be more readily obtained from the resources of the said persons than from the resources of the Authority.

Section 3 – Supplementary powers of Fire Authorities

(1) The powers of a Fire Authority shall include power:

(e) To employ the fire brigade maintained by them, or use any equipment so maintained, for purposes other than firefighting purposes for which it appears to the Authority to be suitable and, if they think fit, to make such change as they may determine for any services rendered in the course of such employment or use.

Section 13 – Duty of Fire Authorities to ensure supply of water for firefighting.

A Fire Authority shall take all reasonable measures for ensuring the provision of an adequate supply of water, and for securing that it will be available for use, in case of fire.

Section 30 – Powers of Firemen (Firefighters) and Police in Extinguishing Fires.

(1) Any member of a fire brigade maintained in pursuance of this Act who is on duty, any member of any other fire brigade who is acting in pursuance of any arrangements made under this Act, or any constable, may enter and if necessary break into any premises or place in which a fire has or is reasonably believed to have broken out, or any premises or place which it is necessary to enter for the purposes of extinguishing a fire or of protecting the premises or place from acts done for firefighting purposes, without the consent of the owner or occupier thereof, and may do all such things as he may deem necessary for extinguishing the fire or for protecting from fire, or from acts done as aforesaid, any such premises or place or for rescuing any person or property therein.

(3) At any fire the senior fire brigade officer present shall have the sole charge and control of all operations for the extinction of the fire, including the fixing of the positions of fire engines and apparatus, the attaching of hose to any water pipes or the use of any water supply, and the selection of the parts of the premises, object or place where the fire is, or of adjoining premises, objects or places, against which the water is to be directed.

(4) Any water undertakers shall, on being required by any such senior officer as is mentioned in the last preceding subsection to provide a greater supply and pressure of water for extinguishing a fire, take all necessary steps to enable them to comply with such requirement and may for that purpose shut off the water from the mains and pipes in any area; and no authority or person shall be liable to any penalty or claim by reason of the interruption of the supply of water occasioned only by compliance of the water undertakers with such a requirement.

(5) The senior officer of police present at any fire, or in the absence of any officer of police the senior fire brigade officer present, may close to traffic any street or may stop or regulate the traffic in any street whenever in the opinion of that officer it is necessary or desirable to do so for firefighting purposes.

(6) In this section the expression 'senior fire brigade officer present', in relation to any fire, means the senior officer present of the fire brigade maintained in pursuance of this Act in the area in which the fire originates, or, if any arrangements or reinforcement scheme made under this Act provided that any other person shall have charge of the operations for the extinction of the fire, that other person.

Section 1 (1) (a) The Fire Services Act 1947 (as amended) places a general duty on every fire authority to make provision for fire fighting purposes. Within section 1 (1) (a) the Fire Authority is required to:

"Secure the services for their area of such a fire brigade and such equipment as may be necessary to meet efficiently all normal requirements".

In defining the term 'normal requirements' and its relationship with incident command and legislation there is no statutory definition. However, Halsbury's laws of England make the following observation:

"It is thought that the reference to normal requirements does not imply that a fire brigade

has no obligations in respect of abnormal fires. In considering however, the action to be taken in relation to a fire, regard must be had to all the factors, in particular e.g. the likelihood of other calls on the brigade, the danger to life and property and the value of the property concerned. Thus in certain circumstances, e.g. fire in a refuse dump, it might be best to allow the fire to burn itself out". (4th Edn. Vol. 18. 455, footnote 2)

Whilst this is a general interpretation it clearly identifies that defensive fire fighting may in certain circumstances be a normal requirement at an incident.

A4.2 Health and Safety at Work etc. Act 1974

General duties

The general duties in the 1974 Act relate to all persons at work and the protection of others who might be injured by the activities of persons at work. The aim is to ensure that all persons associated with the workplace, activities, articles and substances take responsibility for health and safety in relation to that workplace.

Employers duties to employees

Employers have responsibility for health and safety of employees.

Section 2 (1) imposes a general duty on the employer to ensure, so far as is reasonably practicable, the health, safety and welfare at work of all his employees.

This is the core provision of the Act and is expanded in section 2 (2). The duty is as concerned with human behaviour as with the physical environment.

Section 2 (2) duties require safe plant, systems of work, substances, training and supervision, safe place of work, access and working environment.

Duties of employer to non employees

Section 3 of the act imposes a general duty on employers for protection of persons other than their own employees. The general duty under 3(1) requires the employer to conduct his undertaking in such a way as to ensure so far as is reasonably practicable, that persons not in his employment who may be affected thereby are not thereby exposed to risks to their health and safety. The section places a duty on employers to have regard to safety of members of the public.

Duties of people in control of premises

Section 4 of the act imposes a duty of care upon the controller (occupier) of premises to ensure, so far as is reasonably practicable that the premises, all means of access thereto or egress therefrom and any plant or substances in the premises, or provided for use there, is safe and without risks to health.

Duties regarding emissions into the atmosphere

Section 5 of the act imposes a general duty on persons in control of certain premises in relation to harmful emissions into the atmosphere. The Environment Act 1995 provides for the control of pollution responsibilities contained in part 1 of the Health and Safety at Work Act 1974 to be transferred to the Environment Agency as from 1st April 1996.

Supply of goods

Section 6 of the Act imposes duties on those manufacturing or supplying articles to the workplace.

Employees duty to take reasonable care

Section 7 of the Act imposes a duty upon each individual employee to take reasonable care while at work for the health and safety of him/herself and other persons.

Section 8 not intentionally or recklessly to interfere

Imposes a duty on all persons not intentionally or recklessly to interfere with or misuse anything provided in the interests of health, safety and welfare under the relevant statutory provisions.

This section can be invoked against anyone, not just employees.

Liability is for interference with something which has been provided specifically to comply with safety legislation.

Directors liability section 37

Section 37 enables liability to be imposed on senior management where the organisation is in breach of its duties. This section may only be invoked against the most senior management, who represent the alter ego (very brain) of the organisation. The section is particularly relevant to senior managers who have omitted to set up a safe system of work or to curb unsafe behaviour.

A4.3 Water Resources Act 1991

Sections 85-89 cover offences relating to polluting controlled waters.

Section 85 identifies contravention where a person causes or knowingly permits poisonous, noxious, polluting or solid matter to enter any controlled water. This includes trade effluent. The section also highlights contravention where a person permits any matter whatever to enter any inland freshwater so as to tend to impede the proper flow of the waters in a manner leading to, or likely to lead to a substantial aggravation of:

(a) pollution due to other causes;

(b) the consequences of such pollution

Section 89 provides defence as follows:

(1) A person shall not be guilty of an offence under section 85 ... if:

(a) the entry is caused or permitted, or the discharge is made in any emergency in order to avoid danger to life or health;

(b) That person takes all steps as are reasonably practicable in the circumstances for minimizing the extent of the entry or discharge and of its polluting effects;

(c) Particulars of the entry or discharge are furnished to the Authority as soon as reasonably practicable after the entry occurs.

A4.4 The Environment Act 1995

Established the Environment Agency who are the authority referred to in the Water Resources Act 1991.

The Environment Agency has legal responsibilities enabling it to take remedial action in respect of controlled waters which have become polluted in an emergency situation, and addressing waste management matters brought about by the emergency situation. The Agency has responsibility relating to defined matters concerning air, land and water which impact the environment. The Agency's responsibility does not extend to the effects of smoke on the environment other than when confined in a site regulated under the environment protection act.

In any action for pollution involving operations on a fire service incident ground, the Fire Authority, through the Incident Commander should be able to demonstrate that it took all practicable steps to minimize the pollution and that the balance at the time was one in which an emergency existed which involved danger to health or life which was greater than allowing discharge to occur.

A4.5 Regulations under the HSWA 1974 Relevant to Incident Command

In 1992 a 'six pack' of Health and Safety regulations were introduced. These regulations made explicit those duties and responsibilities which were implied by the 1974 Act.

The following regulations are relevant to incident command because they are a major shift away from the prior safety legislation which looked at types of workplace or premises. The new regulations concentrate on combating a specific type of risk rather than the premises in which the risk occurred. As Risk assessment and management is a crucial part of incident command relevant factors are considered below.

A4.6 Management of Health and Safety at Work Regs 1992

Regulation 3 – Risk Assesssment.

Every employer shall make a suitable and sufficient assessment of;

(a) The risks to the health and safety of his employees to which they are exposed whilst they are at work;

and

(b) the risks to the health and safety of persons not in his employment arising out of or in connection with the conduct by him, of his undertaking.

For the purposes of identifying the measures he needs to take to comply with the requirements and prohibitions imposed upon him by or under the relevant statutory provisions.

Any assessment such as referred to in (a) or (b) above shall be reviewed by the employer if:

There is reason to suspect that it no longer valid; or

There has been a significant change in the matters to which it relates; and

Where as a result of any such review changes to an assessment are required, the employer shall make them.

Regulation 4 – Health and safety arrangements.

Every employer shall make and give effect to such arrangements as are appropriate, having regard to the nature of his activities and the size of his undertaking, for the effective planning, organisation, control, monitoring and review of the preventative and protective measures.

Regulation 7 – Procedures for serious and imminent danger and for danger areas.

Every employer shall;

(a) Establish and where necessary give effect to appropriate procedures to be followed in the event of serious and imminent danger to persons at work in his undertaking Nominate a sufficient number of competent people to implement those procedures insofar as they relate to the evacuation of the premises of persons at work in his undertaking; and Ensure that none of his employees has access to any area occupied by him to which it is necessary to restrict access on the grounds of health and safety unless the employee has received adequate health and safety instruction.

Regulation 9 – Co-operation and co-ordination.

Where two or more employers share a workplace each such employer shall ;

(a) Co-operate with the other so far as is necessary to comply with the requirements and prohibitions imposed on them by or under the relevant statutory provisions.

(b) (Taking into account the nature of his activities) take all reasonable steps to co-ordinate the measures he takes to comply with the requirements and prohibitions imposed upon him by or under the relevant statutory provisions with the measures the other employers concerned are taking to comply with the requirements and prohibitions imposed on them by or under the relevant statutory provisions.

Take all reasonable steps to inform the other employers concerned of the risks to their health and safety arising out of or in connection with the conduct by him of his undertaking.

Regulation 10 – Persons working in host employers undertaking.

Every employer … shall ensure that the employer of any employees from an outside undertaking who are working on the undertaking is provided with comprehensible information on;

The risks to those employees health and safety arising out of or in connection with the undertaking; and

The measures taken by the main employer to comply with the requirements and prohibitions imposed by or under the relevant statutory provisions.

Every employer shall ensure that any person working in his undertaking who is not his employee is provided with appropriate instructions and comprehensible information regarding any risks to the persons health and safety from the work activities.

Every employer shall also ensure that other employees are aware of emergency arrangements and receive sufficient information to enable them to identify persons nominated to implement evacuation procedures.

A4.7 Workplace (Health Safety and Welfare Regs 1992)

Regulation 5 – (ACOP) Maintenance of workplace, and of equipment, devices and systems.

The workplace, and the equipment and devices mentioned in these regs should be maintained in an efficient state, in efficient working order and in good repair. If a potentially dangerous defect is discovered, the defect should be rectified immediately or steps taken to protect anyone who may be put at risk, for example by preventing access until the work can be carried out or equipment replaced.

A4.8 Provision and use of work equipment Regulations 1992

Regulation 5 – suitability of work equipment.

Every employer shall ensure that work equipment is so constructed or adapted to be suitable for the purpose for which it is to be used or provided.

In selecting work equipment, every employer shall have regard to the working conditions and to the risks to health and safety of persons which exist in the premises or undertaking in which that work equipment is to be used and any additional risks posed by the use of that work equipment.

Every employer shall ensure that work equipment is used only for operations for which, and under which, it is suitable.

Regulation 7 – Specific risks

Where the use of work equipment is likely to involve a specific risk to health and safety, every employer shall ensure that.

The use of that work equipment is restricted to those persons given the task of using it;

The ACOP identifies that regulation 5 addresses the safety of work equipment from three aspects;

(a) its initial integrity

(b) the place where it will be used

(c) the purpose for which it will be used

This requires the employer to assess the location and take into account any risks that may arise from the particular circumstances. Employers should also take into account the fact that work equipment itself can sometimes cause risks to health and safety in particular locations which would otherwise be safe, ie generators in enclosed spaces.

Regulation 12 – protection against specific hazards

One specified hazard particularly relevant to the incident ground is Material falling from equipment, i.e. work at height or involving aerial appliances.

A4.9 Personal Protective Equipment at Work Regulations 1992

Regulation 4 provision of personal protective equipment Every employer shall ensure that suitable PPE is provided to his employees who may be exposed to a risk to their health or safety whilst at work except where and to the extent that such risk has been more adequately controlled by other means equally or more effective.

PPE shall not be suitable unless;

(a) It is appropriate for the risk or risks involved

and the conditions at the place where exposure to the risk may occur.

(b) It takes account of ergonomic requirements and the state of health of the persons who may wear it.

So far as is practicable, it is effective to prevent or adequately control the risk or risks involved without increasing overall risk.

Regulation 6 – Assessment of PPE

The assessment shall include

An assessment of any risk or risks to health or safety which have not been avoided by other means.

Review of the assessment if there is reason to suspect it is no longer valid or there has been significant change in the matters to which it relates, and where as a result of any such review, changes are required, the employer shall make them.

Other regulations relevant to Incident command

It can be seen that regulations made since the HASAWA 1974, are based on the principles of risk assessment. The requirement for risk control measures such as information, instruction (both guidance and authoritative), training and supervision are also common themes. Therefore this subordinate legislation can provide the Fire Officer with guidance on the hazards and risks associated with specific activities and methods of control. Suggested further reading which should be viewed against the domain the incident commander will experience includes;

- FIRE SERVICE HEALTH & SAFETY
  Volume 1: A Guide for Seniors Officers
  Volume 2: A Guide for Managers
  Volume 3: A Guide to Operational Risk Assessment
  Dynamic management of risk at operational incidents
- COSHH 1994
- MANUAL HANDLING REGS 1992
- NOISE AT WORK REGS 1989
- ELECTRICITY AT WORK REGS 1989
- OCCUPIERS LIABILITY ACT 1957 (AS AMENDED BY THE OCCUPIERS LIABILITY ACT 1984)

A4.10 Civil Law

Vicarious liability

The basic principle of vicarious liability is that the employer owes a duty of care to his/her workforce not only in respect of his/her own activities but in addition, in respect of mistakes made by employees, for whom the employer is vicariously, or indirectly liable.

Before vicarious liability is imposed on a dependent there are two conditions to be met.

First, there must be a specific employer-employee relationship. As it can sometimes be difficult to define the status of employment relationships, various tests can be applied. One accepted test is that the person is under a contract of service being an employee rather than a contract for service being independent contractors.

A further test is one of control, the crucial factor being the degree of control exercised by the employer over the way the work is done although, because in many occupations the employer cannot control the way work is done, a wide range of factors are considered by the courts when deciding control.

The second condition which must be met before an employer will be held liable for an employees tort is that it must be committed when the employee is acting in the course of employment. This is very wide ranging and can include acts done during employment even if forbidden by the employer.

A4.11 Negligence

The Incident Commander should be aware of what circumstances could lead to a civil action against the fire authority for negligence. This is an independent tort for which the existence of a duty of care is a pre-requisite for liability.

A Tort is a "wrong", the law of tort is simplistically the civil law of wrongs. The most important component of the law of tort is the tort of negligence. In essence, it requires people should have consideration of others when they act carelessly. Where such carelessness causes damage, loss or injury, monetary compensation should be available.

A precedent set in law in 1932, through the central reasoning by the House of Lords set a clear general principle for negligence. This case set out a general rule entitling those carelessly harmed by another to sue, so long as the harm was reasonably foreseeable. The House of Lords did not confine their comments to the particular facts, the comments made discussed the issue of negligent behavior generally. The case has been referred to and expanded on in numerous cases since.

Donoghue *vs* Stevenson 1932

The plaintiff went to a café with a friend and ordered some drinks. The plaintiff drank some ginger beer, as her friend poured the remainder of the contents into the glass, it contained the remnants of a decomposed snail, causing her to become ill. As the bottle was opaque the retailer could not be accountable so the manufacturer was sued.

Held by the House of Lords

"You must take reasonable care to avoid acts or omissions which you can reasonably foresee would be likely to injure your neighbour" Neighbour, in the legal sense he defined as **"persons who are so closely and directly affected by my act that I ought reasonably to have them in contemplation as being so affected when I am directing my mind to the acts or omissions which are called in question"**.

The test of reasonableness is based on foresight rather than after the event. Whether the care that has been taken is or is not reasonable is a question the answer to which varies with circumstances. Therefore it does not follow that every error of judgment amounts to negligence. The failing must be in the standards of care required in the particular circumstances of the case. Furthermore for liability to succeed Three aspects of the duty of care must exist;

- A duty of care was owed.
- The duty of care has been breached.
- That this resulted in Injury, loss or damage.

Recent cases involving professional negligence actions are relevant to incident commanders in relation to their actions or omissions to act at an incident.

The **Bolam Test** became a measure for judging that the standard of professional behaviour is not that of the ordinary man (what an ordinary reasonable person would consider reasonable).

The defendant is judged by the standard of the ordinary skilled person exercising and professing to have that special skill

In the case, Bolam *vs* Friern Hospital Management Committee 1957, the plaintiff agreed to undergo electro-convulsive therapy during which he suffered a fractured pelvis. The issue was whether the doctor was negligent in failing to give a relaxant drug, or in failing to provide means of restraint. Evidence was given of various practices of various doctors in use of relaxant drugs during E.C.T. treatment.

Held: The action was not successful, in this situation the defendant was not held negligent if he acts in accordance with a practice accepted at the time as proper by a responsible body of professional opinion skilled in the particulars form of treatment.

This test, set down in a high court decision approved by the House of Lords is of general application and not just limited to doctors. The test is not conclusive because a judgment of proximity between plaintiff and defendant needs to be made.

In interpreting the duty of care regarding the fire service **The Capital and Counties case (1997) 2 All ER 865**, the court of appeal considered three cases where the owner or occupier sued brigades for negligence following attendance at fires.

The court decided in what circumstances a fire brigade owed a duty of care. It concluded that a fire brigade did not enter into sufficiently proximate relationship with the occupier merely by attending the incident or fighting the fire. However if the fire brigade by their actions, increased the risk of the danger which had caused the damage to the plaintiff they would be liable for negligence in respect of that damage, unless they could show that the damage would have occurred in any event.

It was accepted that the Bolam test should be applied to fire brigades and indicated that this was a very high threshold in establishing negligence, namely that it must be established that the error was one that no reasonably well informed and competent firefighter could have made.

Reasonably Practicable

This term can be found in civil law and in criminal law relating to health and safety, however what does it mean?

The most significant judicial authority on the meaning of this term is to be found in Edwards *vs* National Coal Board (1949).

"Reasonably practicable is a narrower term than physically possible, and implies that a computation must be made in which the quantum of risk is placed in one scale, and the sacrifice, whether in money, time or trouble, involved in the measures necessary to avert the risk, is placed in the other; and that, if it be shown that there is a gross disproportion between them, the risk being insignificant in relation to the sacrifice, the person on whom the duty is laid discharges the burden of proving that compliance was not reasonably practicable. This computation falls to be made at a point of time anterior to the happening of the incident complained of".

A4.12 The Tort of Breach of Statutory Duty

An injured employee may also sue the employer if he was in breach of a statutory duty unless the relevant legislation prevents (disassociates) this course of action.

Where the employer has been found to breach a statutory duty imposed by legislation, certain regulations may support civil liability for this breach. In such cases, it will be necessary for the plaintiff seeking to rely on tort, breach of statutory duty to prove;

(i) The duty was owed to that plaintiff, either individually or as a member of a class meant by parliament to be protected.

(This is not normally difficult because of the breadth of coverage of the regs).

(ii) The duty was owed to the defendant.

(This is supported by the fact that most regs are not limited to specific workplaces or activities)

(iii) The defendant was in breach of the statutory duty.

The regulations and ACOP'S may give useful support to evidence due to their extensive nature and detailed suggestion. They are admissible in evidence in civil actions. Civil Law cases are viewed on the 'balance of probability', rather than the criminal view of 'beyond reasonable doubt'. It can be seen that failure to comply with the ACOP or provide some action equally suitable, would prove the breach.

(iv) The damage caused was of the type to be prevented by the statute or regulation.

Whilst pre 1974 acts were very prescriptive, post HASAWA legislation covers most workplaces and activities.

APPENDIX 5 – Incident Command System: Further Examples of Application

The following diagrams are further examples of how the Incident Command System structure may be applied to incidents.

Figure A 5/1 Example of Sector Development during an escalating incident.

Figure A 5/2 Example of Sector Designation, multiple road traffic accident.

Figure A 5/3 Example of Sector Designation, Civil Disturbances.

Figure A 5/4 Example of Tactical mode, Hazardous substance release, Defensive mode.

Figure A 5/5 Example of Tactical mode, Hazardous substance release, Offensive mode.

Figure A 5/6 Example of Tactical mode, Hazardous substance release, Transitional mode.

Figure A 5/7 Example of Tactical mode, multiple road traffic accident, Transitional mode.

APPENDIX 5

Figure A 5/1

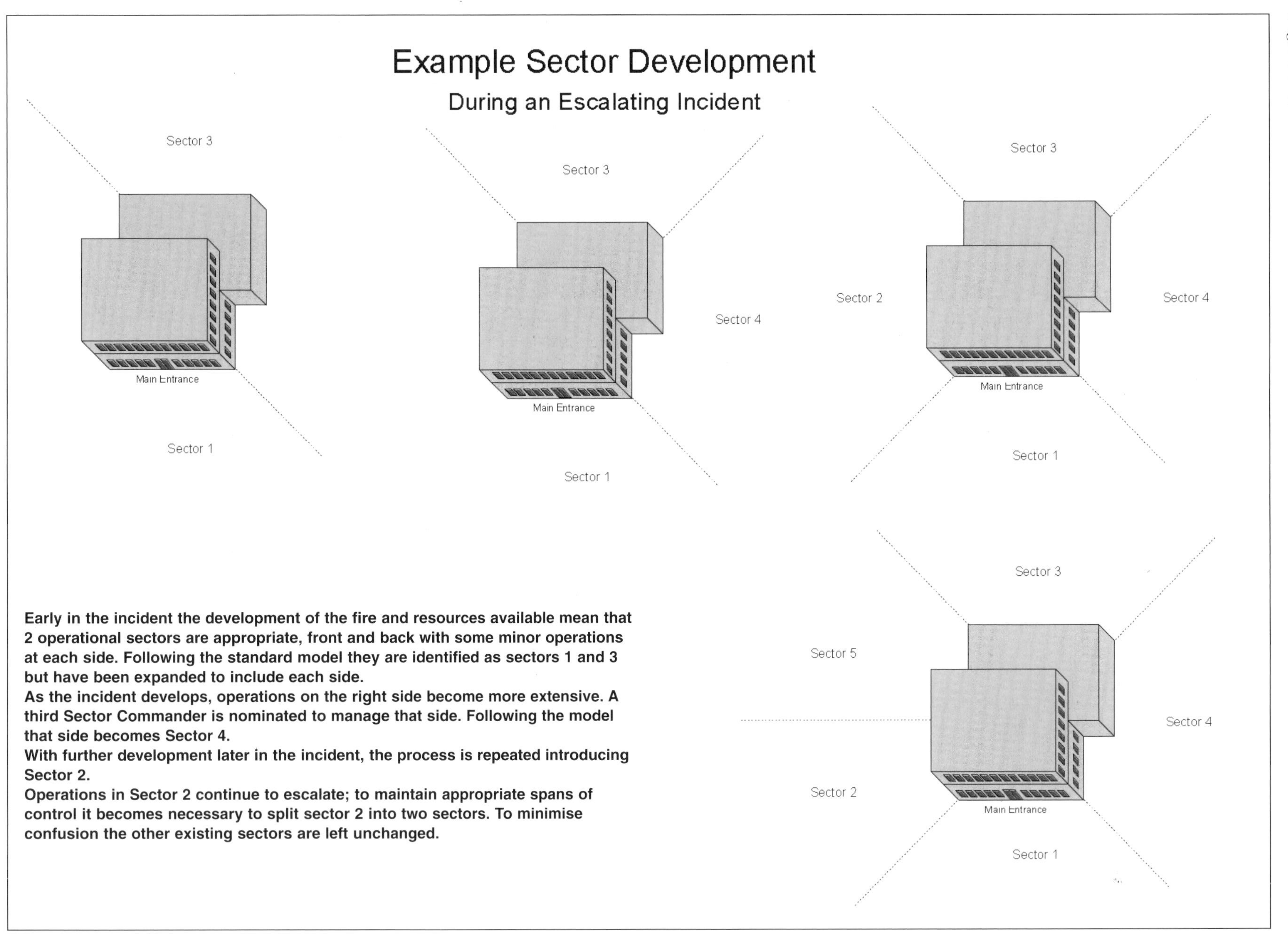

Figure A 5/2

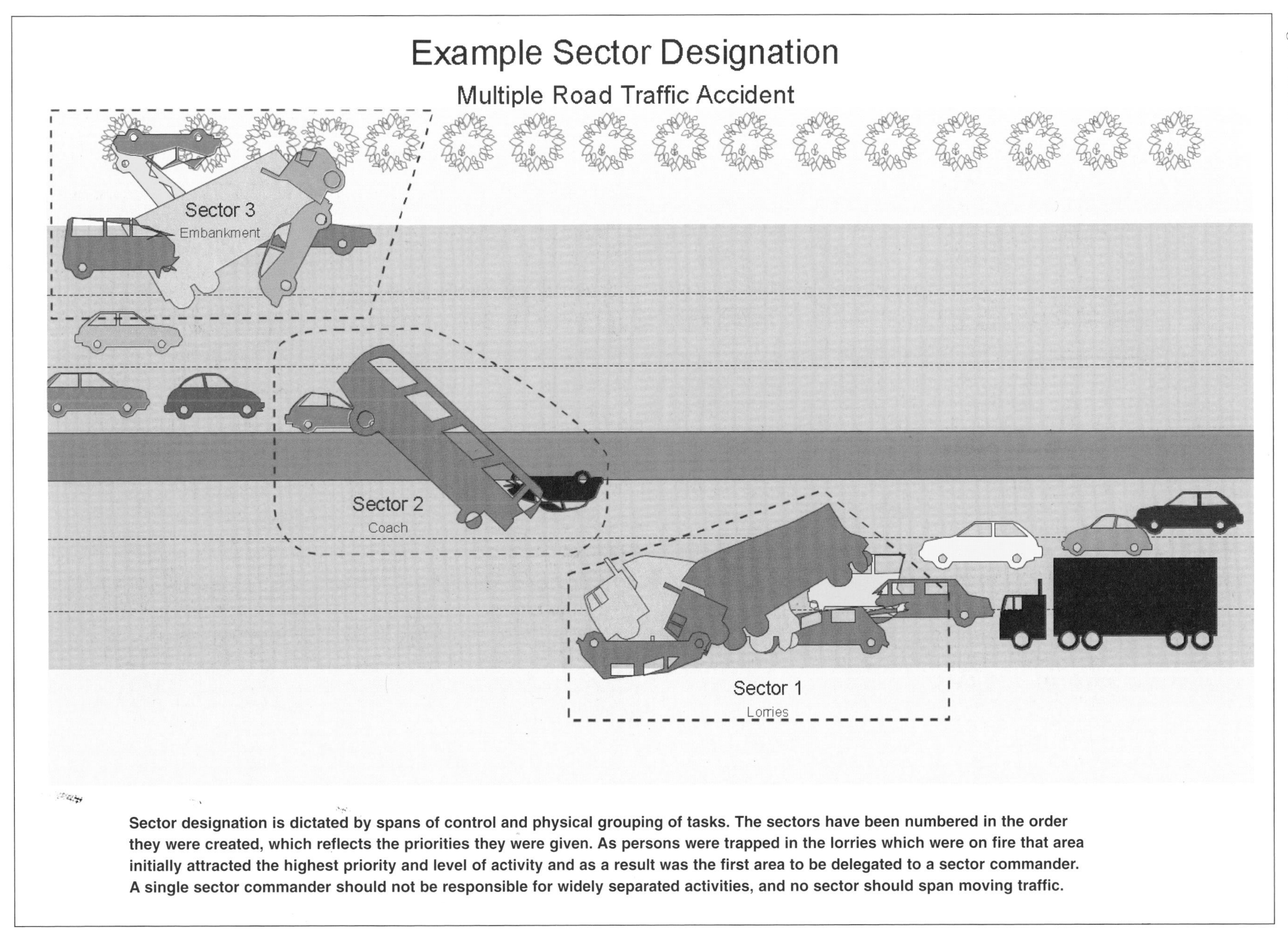
Example Sector Designation
Multiple Road Traffic Accident
Sector 3
Embankment
Sector 2
Coach
Sector 1
Lorries
Sector designation is dictated by spans of control and physical grouping of tasks. The sectors have been numbered in the order they were created, which reflects the priorities they were given. As persons were trapped in the lorries which were on fire that area initially attracted the highest priority and level of activity and as a result was the first area to be delegated to a sector commander. A single sector commander should not be responsible for widely separated activities, and no sector should span moving traffic.

Figure A 5/3

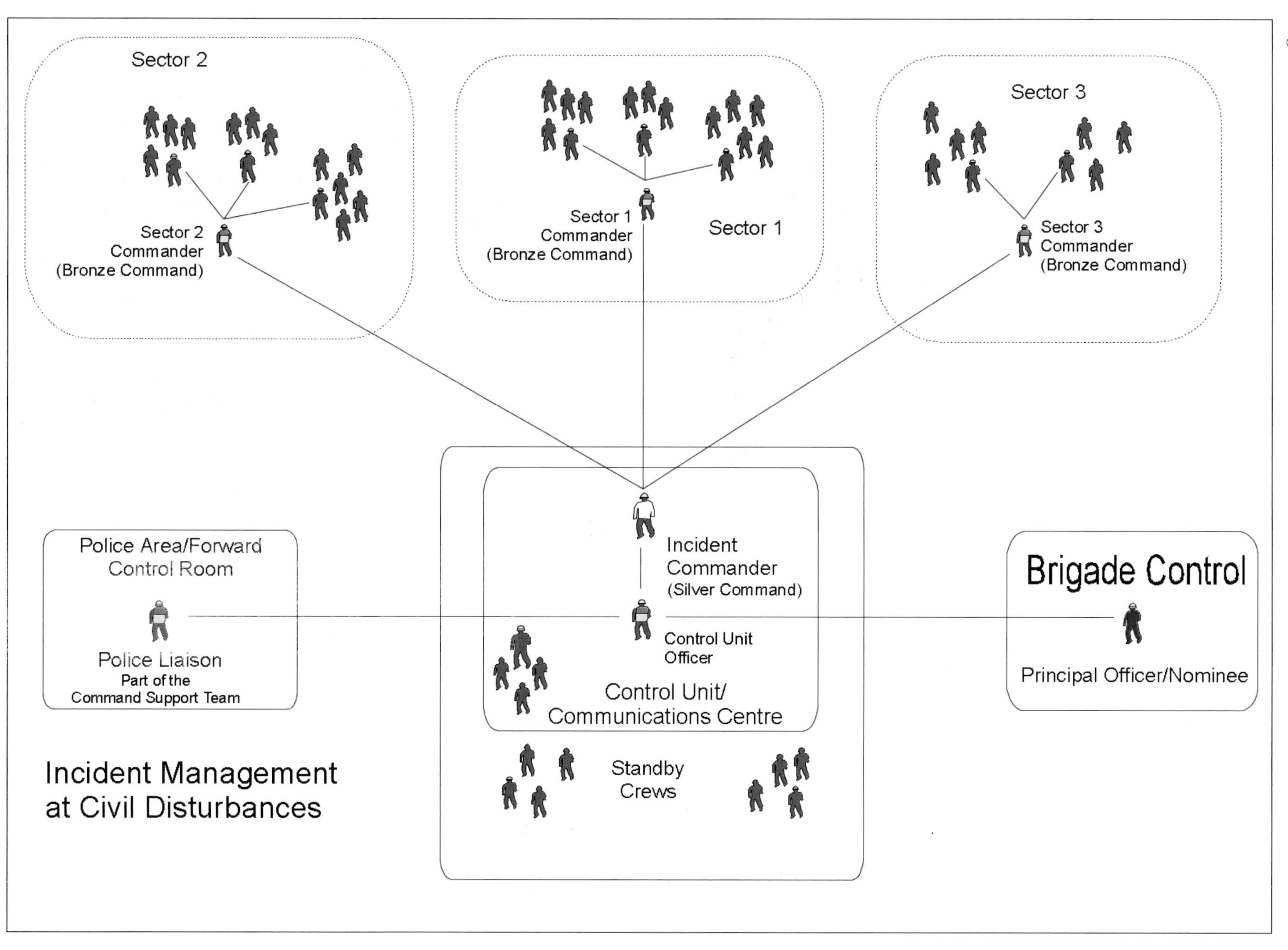

Sector 2
Sector 2 Commander (Bronze Command)
Sector 1
Sector 1 Commander (Bronze Command)
Sector 3
Sector 3 Commander (Bronze Command)
Police Area/Forward Control Room
Police Liaison
Part of the Command Support Team
Incident Commander (Silver Command)
Control Unit Officer
Control Unit/ Communications Centre
Standby Crews
Brigade Control
Principal Officer/Nominee
Incident Management at Civil Disturbances

Figure A 5/4

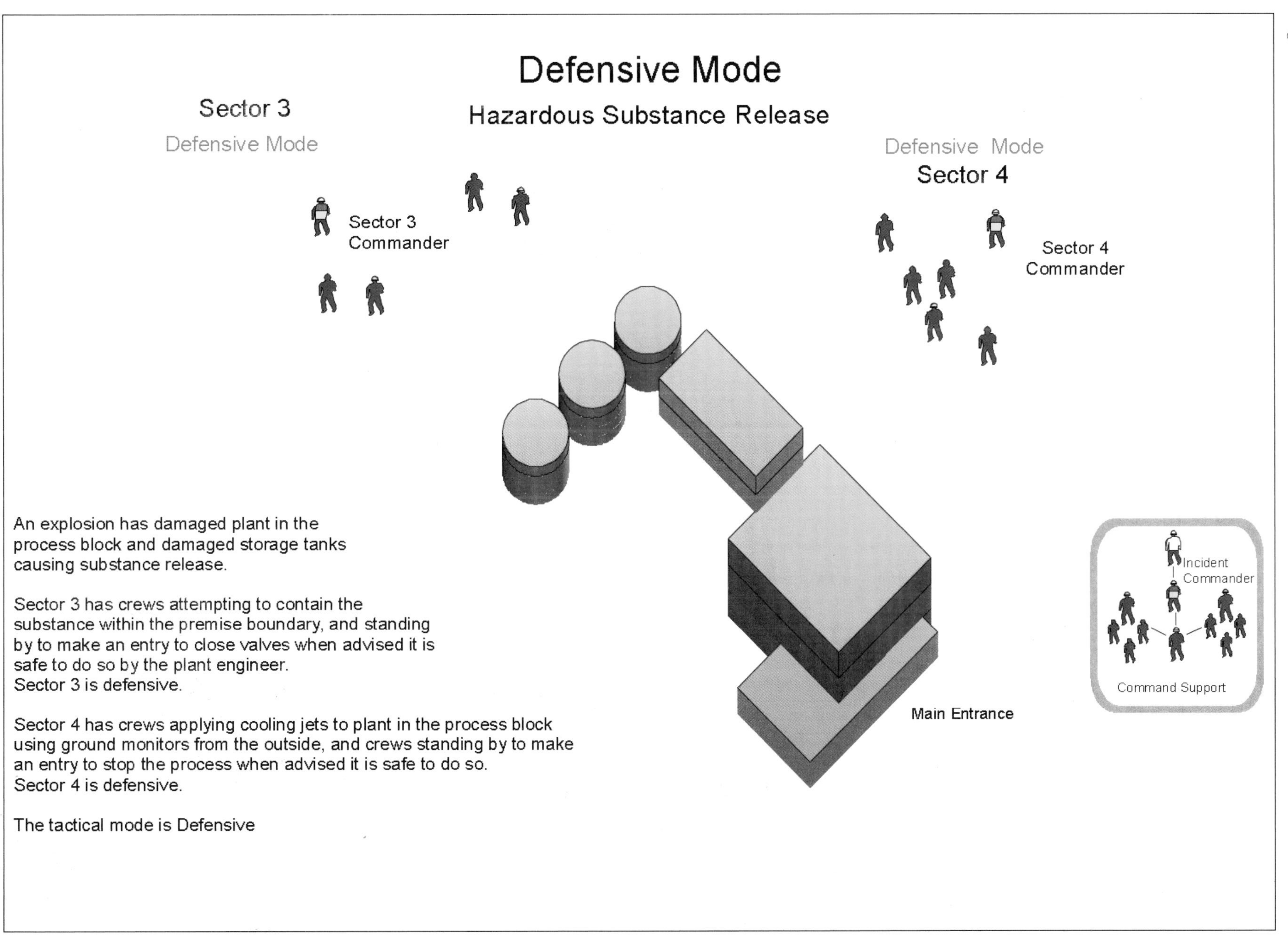
Defensive Mode
Hazardous Substance Release
Sector 3
Defensive Mode
Sector 3
Commander
Defensive Mode
Sector 4
Sector 4
Commander
Main Entrance
Incident
Commander
Command Support
An explosion has damaged plant in the
process block and damaged storage tanks
causing substance release.
Sector 3 has crews attempting to contain the
substance within the premise boundary, and standing
by to make an entry to close valves when advised it is
safe to do so by the plant engineer.
Sector 3 is defensive.
Sector 4 has crews applying cooling jets to plant in the process block
using ground monitors from the outside, and crews standing by to make
an entry to stop the process when advised it is safe to do so.
Sector 4 is defensive.
The tactical mode is Defensive

Figure A 5/5

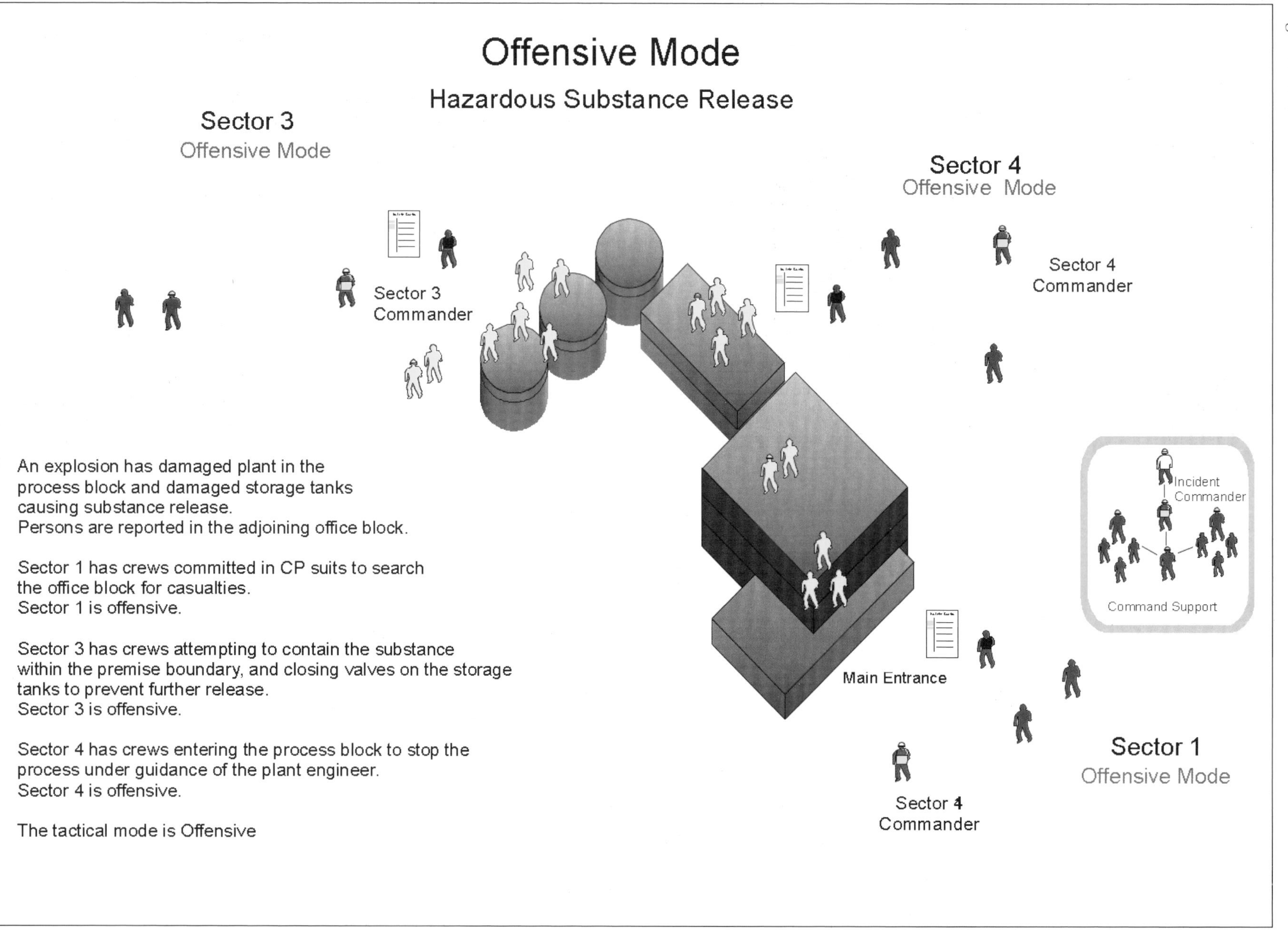
Offensive Mode
Hazardous Substance Release
Sector 3
Offensive Mode
Sector 3
Commander
Sector 4
Offensive Mode
Sector 4
Commander
Incident
Commander
Command Support
Main Entrance
Sector 4
Commander
Sector 1
Offensive Mode
An explosion has damaged plant in the process block and damaged storage tanks causing substance release.
Persons are reported in the adjoining office block.
Sector 1 has crews committed in CP suits to search the office block for casualties.
Sector 1 is offensive.
Sector 3 has crews attempting to contain the substance within the premise boundary, and closing valves on the storage tanks to prevent further release.
Sector 3 is offensive.
Sector 4 has crews entering the process block to stop the process under guidance of the plant engineer.
Sector 4 is offensive.
The tactical mode is Offensive

Figure A 5/6

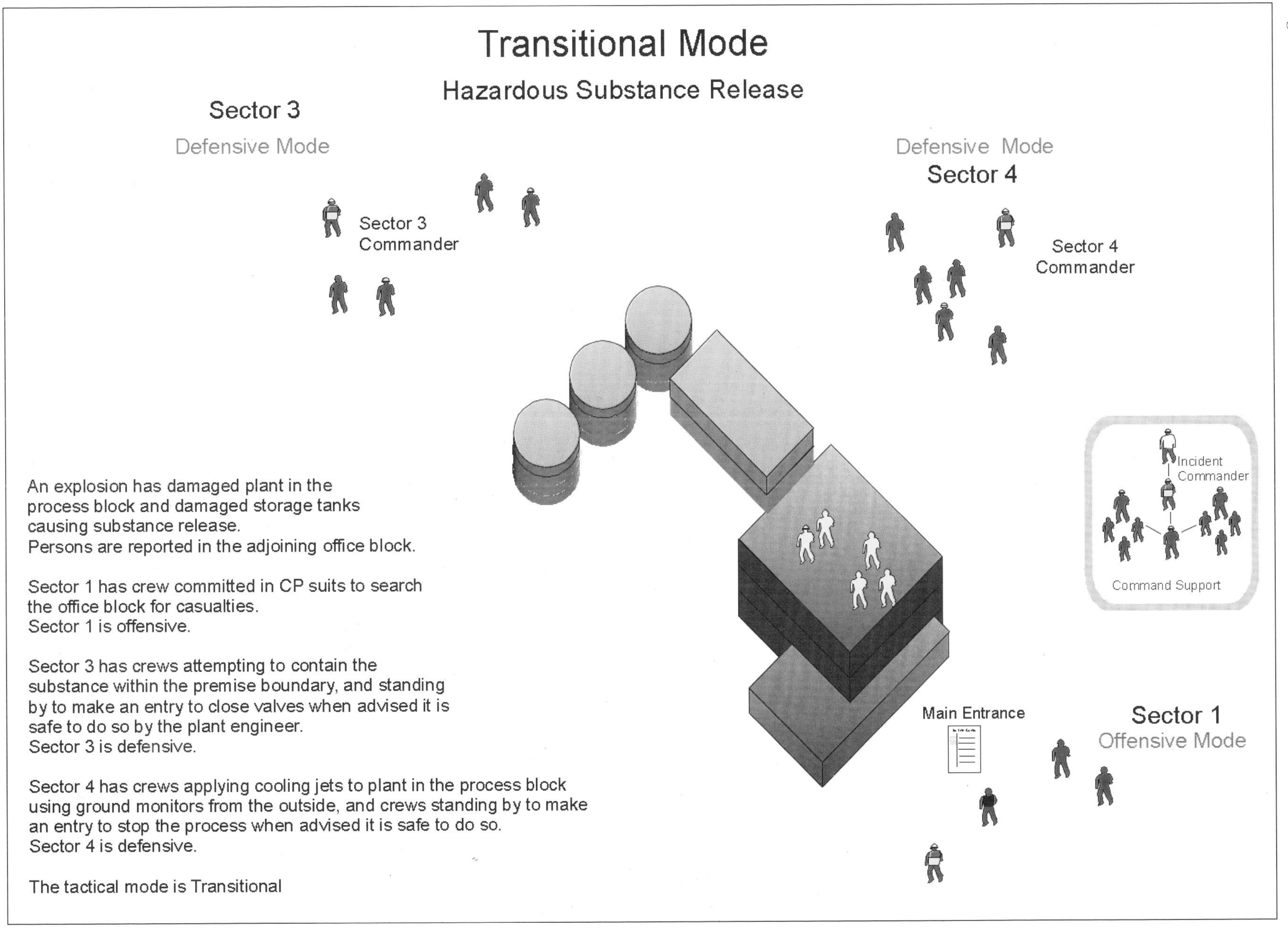
Transitional Mode
Hazardous Substance Release
Sector 3
Defensive Mode
Sector 3
Commander
Defensive Mode
Sector 4
Sector 4
Commander
Incident
Commander
Command Support
Main Entrance
Sector 1
Offensive Mode
An explosion has damaged plant in the process block and damaged storage tanks causing substance release.
Persons are reported in the adjoining office block.
Sector 1 has crew committed in CP suits to search the office block for casualties.
Sector 1 is offensive.
Sector 3 has crews attempting to contain the substance within the premise boundary, and standing by to make an entry to close valves when advised it is safe to do so by the plant engineer.
Sector 3 is defensive.
Sector 4 has crews applying cooling jets to plant in the process block using ground monitors from the outside, and crews standing by to make an entry to stop the process when advised it is safe to do so.
Sector 4 is defensive.
The tactical mode is Transitional

Figure A 5/7

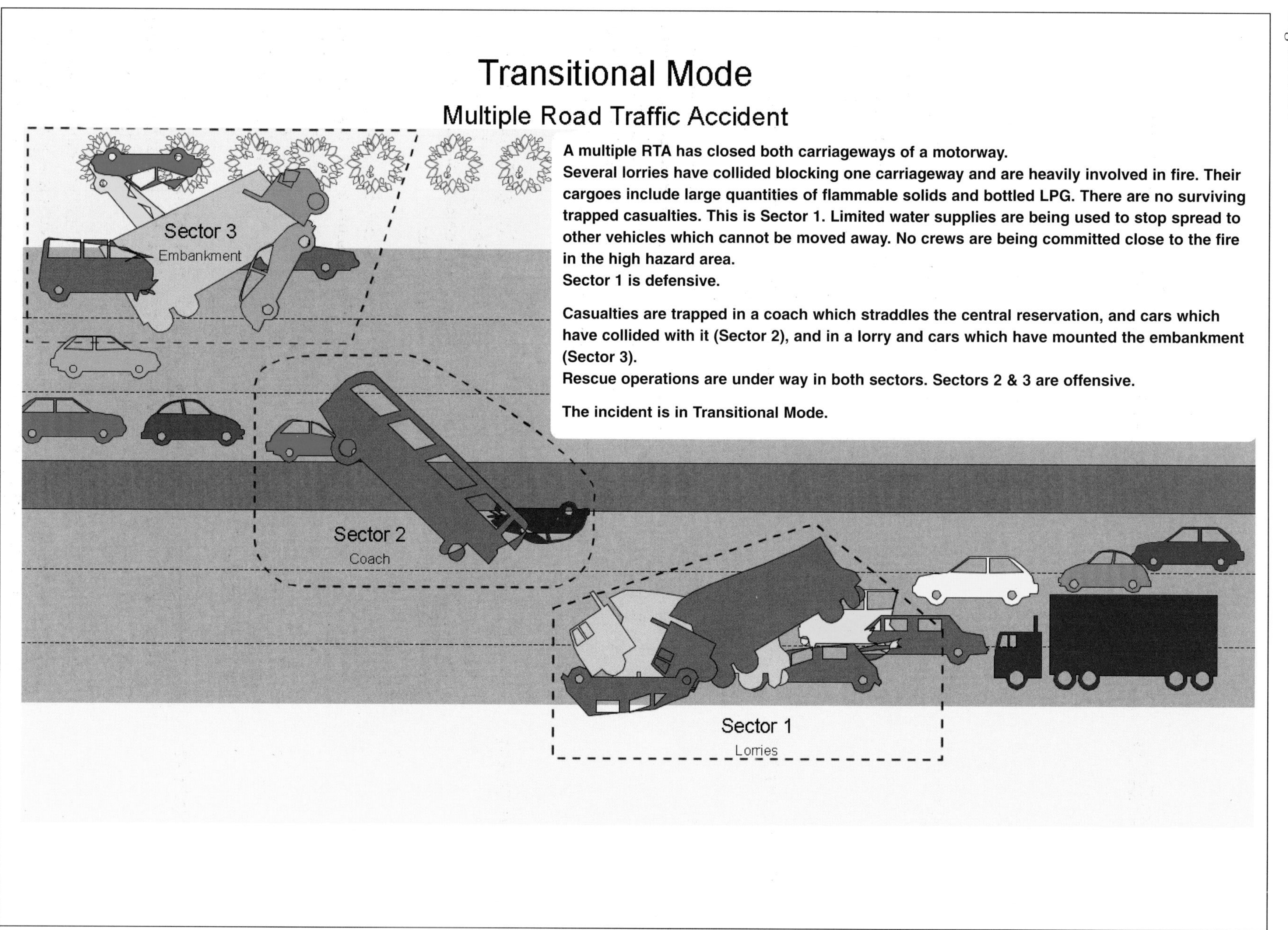
Transitional Mode
Multiple Road Traffic Accident
A multiple RTA has closed both carriageways of a motorway.
Several lorries have collided blocking one carriageway and are heavily involved in fire. Their cargoes include large quantities of flammable solids and bottled LPG. There are no surviving trapped casualties. This is Sector 1. Limited water supplies are being used to stop spread to other vehicles which cannot be moved away. No crews are being committed close to the fire in the high hazard area.
Sector 1 is defensive.
Casualties are trapped in a coach which straddles the central reservation, and cars which have collided with it (Sector 2), and in a lorry and cars which have mounted the embankment (Sector 3).
Rescue operations are under way in both sectors. Sectors 2 & 3 are offensive.
The incident is in Transitional Mode.
Sector 3
Embankment
Sector 2
Coach
Sector 1
Lorries

Glossary of Terms: Incident Command

COMMAND The authority for an agency to direct the actions of its own resources (both personnel and equipment).

COMMAND LINE The line or chain of command at an incident. The ICS relies upon a single unified command line. With the exception of urgent safety related issues officers should not take control of operations outside their assigned responsibility and should ensure all information and instruction is passed via the relevant command line officers.

COMMAND POINT Point from which Incident Commander operates, this may be a car, appliance, specialist unit or part of a building.

COMMAND SUPPORT Command Support is a role undertaken by one or more staff at an incident. The role typically provides recording, liaison, detailed resource management and information gathering for the Incident Commander. At large incidents Command Support may comprise a dedicated team working from a mobile command unit and may include individuals tasked with supporting Sector Commanders.

COMMAND TEAM The ICS relies on shared responsibility and authority. While the Incident Commander retains overall responsibility for the incident, and dictates the overall tactical plan, the decision making for, and control of, local operations is kept as close as possible to those operations.

This is achieved by the creation of a single command line from the crew commander to the Incident Commander. This command line, together with staff tasked with supporting commanders, is the Command Team. For ICS purposes the Command Team is usually taken to be the Incident Commander, Operations Commander(s) and Sector Commanders, together with Command Support staff.

CONTACT POINT A designated point (usually an appliance not involved in operations or an officer's car) from which a nominated Junior Officer or Firefighter will carry out the Command Support function at a small to medium size incident.

CONTROL The authority to direct strategic and tactical operations in order to complete an assigned function and includes the ability to direct the activities of other agencies engaged in the completion of that function. The control of an assigned function also carries with it a responsibility for the health and safety of those involved.

CREW COMMANDER

An officer or firefighter tasked with supervising specific tasks or meeting specific objectives utilising one or more firefighters.

DYNAMIC RISK ASSESSMENT

The continuous assessment of risk in the rapidly changing circumstances of an operational incident, in order to implement the control measures necessary to ensure an acceptable level of safety. Dynamic Risk Assessment is only appropriate during the time critical phase of an incident which is usually typical of the arrival and escalation phase of an incident. At the earliest opportunity the Dynamic Risk Assessment should be supported by a more analytical risk assessment.

FORWARD COMMAND POST

Point, near the scene of operations, where the officer delegated responsibility for command in that area is sited.

HIGH RISE

Incidents are normally sectored on a floor by floor basis, using floor numbers.

INCIDENT COMMANDER

The officer having overall responsibility for dictating tactics and resource management. Overall responsibility for a fire remains with the senior fire brigade officer present under the Fire Services Act Section 30(3), but subject to this a more junior officer may retain the role of Incident Commander.

INNER CORDON

A secured area which surrounds the immediate site of the incident and provides security for it. Such an area will typically have some formal means of access control.

MARSHALLING AREA

Area to which resources not immediately required at the scene or being held for future use can be directed to standby

OPERATIONS COMMANDER

An officer tasked with co-ordinating the operations of several sectors. Responsible directly to the Incident Commander. When an Operations Commander is assigned, operational Sector Commanders will report to the Operations Commander rather than the Incident Commander. Assigning an Operations Commander at an incident which has several operational sectors keeps the span of control of the Incident Commander to be maintained at a satisfactory level.

A newly arriving senior officer mobilised to take command of the incident may elect to assign the current Incident Commander as Operations Commander in order to make best use of that officers knowledge. The demands on the Command Team at incidents attracting less than, say, eight operational pumps are unlikely to warrant the creation of an Operations Commander. It may be more effective to appoint the previous Incident Commander as assistant.

OUTER CORDON	An area which surrounds the inner cordon and seals off a wider area of the incident from the public.
RENDEZVOUS POINT	Point to which all resources at the scene are directed for logging, briefing and deployment.
RISK ASSESSMENT (ANALYTICAL)	A detailed assessment of risks to crews, the public and the environment, based upon as much information as possible. The analytical process should be undertaken as soon as resources permit. Because of the significant amount of time and work involved in carrying out the assessment it may be undertaken by a supporting member of the Command Team rather than a member of the command line. The process is a continuing one so that reviews are completed at appropriate intervals. Note: Unexpected or rapidly developing events may require continued dynamic assessment.
SAFETY	A state where exposure to hazards has been controlled to an acceptable level.
SAFETY OFFICER	Officer delegated specific responsibility for monitoring operations and ensuring safety of personnel working on the incident ground or a designated section of it.
SAFE SYSTEMS OF WORK	A formal procedure which results from systematic examination of a task in order to identify all the hazards and risks posed. It defines safe methods to ensure that hazards are eliminated or risks controlled as far as possible.
SECTOR	A sector is the area of responsibility of a Sector Commander. (ie. a sector should not be created unless someone is given the responsibility for running it.) Sectors should be created to allow effective delegation of command responsibility and authority when an incident is too complex, or too wide spread, to be managed by a single individual. Boundaries between geographic sectors may be geographic features, walls, roads, differences in elevation or separate areas of plant. Ideally, the boundary should be intuitively obvious. Sector size should be dictated by the ability of the Sector Commander to exercise effective control over activities within the sector. If a sector becomes too large or the operations within it too complex for a single person to manage, it should be split into two or more sectors, each with a Sector Commander. Operational sectors are those dealing directly with the incident, typically operational sectors will undertake fire fighting, rescue, cooling and so on. Operational sectors tend to place greater demands on the Sector Commander and demand a smaller span of control than support sectors.

Operational Sectors are usually identified by numbers.

Support sectors are those not dealing directly with the incident. Support sectors are usually defined by the function they undertake, for instance decontamination, foam supply, marshalling or water supply. They may be less dynamic than operational sectors, in which case they may be managed with a greater span of control.

SECTOR COMMANDER

An officer tasked with responsibility for tactical and safety management of a clearly identified part of an incident. Subject to objectives set by the Incident Commander the Sector Commander has control of all operations within the sector, and must remain within it.

The Sector Commander is responsible for all aspects of safety within the sector and may assign safety officers as appropriate. All communications between the sector and other sectors, or other members of the Command Team must go via the Sector Commander. Where appropriate command support may be utilised by the Sector Commander.

SPAN OF CONTROL

The number of people who must have an officer's attention for briefing, reporting, passing instructions or other incident management concerns, in order to carry out their role at the incident.

As a guide five such reporting lines are considered the usual maximum for an Incident Commander to maintain during an incident. This may be increased at an incident which is well in hand or have to be reduced to two or three at a rapidly escalating or highly complex incident. Management of the Span of Control must be effective throughout the command line.

Span of Control concerns are most relevant in circumstances where there are frequent or high volume communications rather than sporadic one-off contacts. A stream of such one-off contacts from many different individuals can be as demanding, or even more so, than continuous contact with a single individual. This should be anticipated and managed appropriately.

STRATEGY

Plans formulated by a Brigade to deal with incidents occurring in its area. The application of the strategy to the incident, after the initial assessment, in order to plan and direct the incident.

TACTICS

The deployment of personnel and equipment on the incident ground to achieve the aims of the strategic plan.

Bibliography

Fire Services Act 1947. HMSO, London.

Fire Services Act 1959. HMSO, London.

Health and Safety at Work Act 1974. HMSO, London.

Control of Substances Hazardous to Health Regulations 1988. Approved Code of Practice. HMSO, London.

HSE (1984) *Training for Hazardous Occupations: A Case Study of the Fire Service.* HMSO, London.

Health and Safety, *a fire service guide – Dynamic management of risk at operational incidents.* (1998) TSO, London.

EFSLB (1995) *Emergency Fire Service Supervision and Command NVQ Level 3 Standards.* HMSO, London.

Principles of Health and Safety at Work, Allan St.John Holt, IOSH publishing.

CACFOA (1994) *Fire Service Major Incident Emergency Procedures Manual.* CACFOA.

National Fire Protection Association (NFPA) (1990) *Fire Department Incident Management System.* Quincy, Mass.

Adair, J. (1968) *Training for Leadership.* McDonald.

Adamson, A. (1970) *The Effective Leader.* Pitman.

Bonney, J. (1995) *Fire command teams: what makes for effective performance?* Fire Service College, Brigade Command Course Project 2/95.

Brunacini, A. (1985) *Fire Command.* Quincy, Mass. National Fire Protection Association.

Charlton, D. (1992, April) *Training and assessing submarine commanders on the Perishers' course.* In collected papers of the First Offshore Installation Management Conference: Emergency Command Responsibilities. Robert Gordon University, Aberdeen.

Cannon-Bowers, J., Tannenbaum, S., Salas, E. & Volpe, C. (1995) *Defining Competencies and establishing team training requirements.* In R. Guzzo & E. Salas (Eds.) *Team Effectiveness and Decision Making in Organisations.* San Francisco: Jossey Bass.

Flin, R. (1996) *Sitting in the Hot Seat: Leaders and Teams for Critical Incident Management.* Wiley, Chichester.

Keampf, G. & Militello, L. (1992) *The Problem of Decision Making in Emergencies.* Fire International No 135, p 38–39.

Klein, R.A. (1995) *National Guidance on Risk Assessment for the UK Fire Service.* Home Office, London.

Kerstholt, J.H. (1997) Dynamic decision making in non-routine situations, in R.Flin, E Salas, M. Strub, & L. Martin, *Decision making under stress.* Ashgate, Aldershot, UK.

Orasanu, J. (1995) Training for aviation decision making: the naturalistic decision making perspective. *Proceedings of the Human Factors and Ergonomics Society 39th annual Meeting.* San Diego, Santa Monica CA: The Human Factors and Ergonomics Society.

Wynne, D. (1995) *Expert teams performing in natural environments.* Fire Service College, Brigade Command Course Project 1/95.

Acknowledgements

HM Fire Service Inspectorate is indebted to the members of the 'Incident Command' Project Group (ICPG) and all who helped with the provision of information and expertise to assist the production of this manual, in particular:

Aberdeen University:	Professor Rhona Flin
CACFOA:	Chief Fire Officer Eric Clark
FBU:	Assistant General Secretary Mike Fordham
FOA:	President Alan Ellis
H.O. Fire Research & Development Group:	Ms Sue Coles
Fire Service College :	Deputy Commandant ACO Paul Grimshaw Senior Divisional Officer Frank Watt
HMIFS: Chairman ICPG:	HM Senior Inspector Peter Morphew HM Inspector Ted Pearn HM Assistant Inspector Tony Boyer
London Fire Brigade	Assistant Chief Officer Charles Hendry Assistant Chief Officer Malcolm Kelly
Northumberland Fire & Rescue Service:	Divisional Officer Alan Locke
West Midlands Fire Service	Assistant Divisional Officer Brian McCutcheon
West Yorkshire Fire Service:	Assistant Chief Officer Kevin Arbuthnot

Photographs:
- Buckinghamshire Fire and Rescue Service
- Fire Service College
- London Fire Brigade
- West Yorkshire Fire Service